SECOND THOUGHTS

Published by
DAWSONS OF PALL MALL
16 Pall Mall, London, S.W.1.
1968
SBN 7129 0221 X

Notes, Commentaries and Transcriptions
© M. A. Hoskin 1968

PRINTED IN GREAT BRITAIN BY
WARREN AND SON LTD.
THE WYKEHAM PRESS
WINCHESTER

THOMAS WRIGHT
OF
DURHAM

SECOND
OR SINGULAR
THOUGHTS
UPON THE
THEORY OF THE UNIVERSE

Edited from the manuscript by
M. A. HOSKIN
Fellow of St. Edmund's House and Lecturer in the
History of Science, Cambridge University

1968
DAWSONS OF PALL MALL
LONDON

No. 10.

Second, or Singular Thoughts
upon the

Theory of the Universe

Design'd as a Supplement
or Sequel to that Already,
Founded upon New Observations
and Discoveries by the Author

In 3 Letters.

Be not Ignorant of any thing Great or Small.
Ecclesiasticus
5: 15

Frontispiece. Original title-page of Wright's MS.

CONTENTS

Editor's Introduction	7
Second Thoughts	17
Preface	19
Letter I	25
Letter II	45
Letter III	61
Epigenoma	75
Index to the Plates	89
Index of Names	91

ILLUSTRATIONS

Wright's Title-page	Frontispiece
Preface, p. 8	facing p. 22
Letter I, p. 28	23
Plate III	40
Plate XXXI of *An Original Theory*	49

EDITOR'S INTRODUCTION

In 1755, Thomas Wright, 'then of the City of London, Gent', purchased from his brother John the house at Byers Green, near Durham, where they had lived as children.[1] Thomas Wright of Durham was then in his mid-forties.[2] Of humble origins, he had used his natural talents and capacity for hard work to bring himself to the notice of the aristocracy, and for over twenty years he had lived on the fringe of high society, surveying estates, giving lessons in mathematics and the physical sciences to noble ladies, and generally making himself useful. In return he had secured a degree of comfort and social position that would otherwise have been denied him. He had published a number of books and minor pieces,[3] on astronomy, architecture, and antiquities, but of these only *Louthiana*,[4] on the antiquities of the Irish county of Louth, could be described as a success. Even so, its sequel was never to reach publication. Wright's major work of scientific interest, *An Original Theory, or New Hypothesis of the Universe*, which had appeared in 1750,[5] had aroused negligible response in England, and efforts that had been made on the Continent to produce a German translation had come to nothing.[6] But now Wright's patrons were dying off, and his thoughts were set on a comfortable retirement in his native village to 'prosicute my Studies'.[7]

Wright still had over thirty years of life remaining to him.[8] Contemporary accounts by visitors to Byers Green or by Wright himself allow us glimpses of his way of life during this period,[9]

---

[1] Edward Hughes, 'The early journal of Thomas Wright of Durham', *Annals of Science* 7 (1951) 1–24, p. 2. This journal is the major source of the first half of Wright's life.

[2] He was born 22 September 1711.

[3] A convenient list of Wright's publications is given in F. A. Paneth, *Chemistry and Beyond*, edited by Herbert Dingle and G. R. Martin with the assistance of Eva Paneth (New York, 1964), pp. 114–16; *Pannauticon*, however, is not a book but an instrument (see the present writer's introduction to the facsimile of Wright's *Original Theory* (in press)). Paneth's bibliography of writings on Wright (*ibid.*, pp. 117–19) is also of value.

[4] *Louthiana* (London, $_1$1748, $_2$1758).

[5] Published in London. A very inferior reprint of the text but without the fine plates was issued with notes by C. S. Rafinesque (Philadelphia, 1837).

[6] This attempt to produce a German translation at Nürnberg is reported by J. H. Lambert (W. Hastie, *Kant's Cosmogony* (Glasgow, 1900), p. lxx).

[7] Hughes, *op. cit.*, p. 2.

[8] He died 25 February 1786.

[9] References are assembled by Paneth, *op. cit.*, pp. 96, 117–18.

but of his 'studies' very little was known until after 19 July 1966, when Messrs. Sotheby auctioned in London some 800 folios of Wright manuscripts, in chaotic condition. These were purchased by Messrs. Dawsons of Pall Mall, who invited the present writer to assist with the task of sorting them. As the pieces of the various jigsaws gradually fell into place, the major item to emerge was a wholly-unsuspected sequel to the *Original Theory*, entitled *Second or Singular Thoughts upon the Theory of the Universe*. The carelessly-written manuscript had been pieced together from numerous fragments, with frequent amendments and repaginations, sometime after 1771,[10] and had grown until it matched the *Original Theory* in length. Seen from the modern standpoint, these *Second Thoughts* are wholly retrograde: they involve a return to the solid heavens of medieval astronomy, and the abandonment of Wright's famous explanation in the *Original Theory* of the Milky Way as an optical effect due to our immersion in a layer of stars. But seen within the context of Wright's lifework, they mark the culmination of his efforts to reconcile the moral and the physical view of the world.

In 1734, as a young man, Wright had argued[11] that the world has a centre in the moral order, the Sacred Throne, and that this is also the centre in the physical order and source of the laws of nature; about this centre the creation is arrayed in spherical symmetry, and comprises an inner sphere or Heaven, which is surrounded by a spherical shell containing the Sun and all the multitudes of stars, beyond which is the darkness of Hell extending outwards indefinitely. In 1750, in the *Original Theory*, Wright had amplified this picture of a single, simple system of stars to allow of innumerable systems, some of them compound, and each encircling one of the innumerable supernatural centres. By this time, too, he had become fascinated with the particular problem of explaining the Milky Way, at which he had made an unsatisfactory first attempt in 1734.[12] He saw that an observer immersed in what approximated *locally* to a layer of stars would observe a

---

[10] Of the items in the manuscript to which a date can be put, the latest are the mention (Letter II, p. 29) of *The Nautical Almanac for the year 1773*, published in 1771, and the reference (*ibid.*, p. 31) to the results of observations of the two transits of Venus, the second of which took place in 1769.

[11] In 'Elements of Existence, or a Theory of the Universe', in Vol. VII of the Wright MSS in the Central Library, Newcastle upon Tyne. Attention was first drawn to this document by Dr. D. M. Knight. It is edited by the present writer for inclusion with the facsimile of Wright's *Original Theory* (in press).

[12] 'Elements of Existence', p. C4.

Milky Way effect as he looked about him along the layer, and Wright reconciled this with his requirements of symmetry about a supernatural centre in two different ways: either the star system formed a vast thin spherical shell about the supernatural centre, in which case the Milky Way would lie roughly in the plane tangent to the shell at the position of the observer; or the star system was flat and ring-shaped, surrounding the supernatural centre as the particles of Saturn's rings surround the planet, in which case the *visible* stars would all lie to one side of the ring.

These explanations, of which Wright is justly proud, had been summarized in 1751 in a Hamburg journal[13] which had chanced to come to the eyes of the young Immanuel Kant.[14] Kant rejected spherical systems on observational grounds, for Maupertuis had observed elliptical nebulae and these Kant believed to be other star systems like our own but viewed obliquely.[15] But he gratefully acknowledged[16] his debt to Wright for the concept of star systems (and ours in particular) as analogous to the solar system, all the stars moving about a common centre and in a common plane as do the planets round the Sun. Now Kant had access neither to Wright's complete text nor to his illustrative plates, and he did not appreciate[17] that for Wright each star system surrounds its own supernatural centre. He therefore assumed that Wright intended the stars of our system to be spread out continuously from one side of the centre (which he thought might be occupied by the star Sirius)[18] across to the other, and he thus inadvertently transformed Wright's flat rings into continuous discs. In so doing he hypnotised generations of readers (including those with access to Wright's original text) into thinking that Wright is to be credited with the modern conception of our Galaxy as a disc-shaped system[19]—a conception which for Wright, as we have seen, was utterly impossible, since he believed our stars must surround a centre in the supernatural order.

Meanwhile, Wright, unaware of the ideas being fathered onto him, was turning over in his mind his attempted reconciliation of the moral and physical view of the world—because, for all its

[13] *Freye Urtheile*, Achtes Jahr (Hamburg, 1751), translated by Hastie, *op. cit.*, Appendix B.
[14] Hastie, *op. cit.*, p. 30.
[15] *Ibid.*, pp. 32–34, 62–64.
[16] *Ibid.*, pp. 30, 54.
[17] *Ibid.*, p. 165.
[18] *Ibid.*, pp. 164–5.
[19] An honourable exception is Vera Gushee in the posthumous ' Thomas Wright of Durham, Astronomer ', *Isis 33* (1941), 197–218.

merits, the conception of the *Original Theory* was imperfect. Some imperfections—its failure to give an adequate explanation of 'the astonishing Phenomenon of several new Stars', or of variable stars, nebulae, and the like—are evident from the text of the *Original Theory*,[20] while others—the absence of a satisfactory account of future rewards and punishments, and the multiplicity of supernatural centres which robbed the universe of any one pre-eminent focal point of the Divine Presence—are best shown by contrast with the 1734 cosmology and with the *Second Thoughts*. But time, Wright believed, is on our side: the passage of events provokes in the observer ideas more fruitful than those arrived at by sheer hard work.[21] And so it was that in 1755, the very year in which Wright began to prepare for retirement, the great earthquake at Lisbon chanced to set him thinking about the construction of the Earth.[22] He was familiar with the views of Halley (and others) that the Earth is not one solid mass, but that 'the External Parts of the Globe may well be reckoned as the Shell, and the internal as a *Nucleus* or inner Globe included within ours, with a fluid medium between';[23] and he imagined that if the fluid medium insinuated itself into the crust until a piece of the crust broke off and sank to the central nucleus, a disturbance would be set up in the fluid: the fluid would splash back and forth but with diminishing force, just as the earthquake shocks recurred at intervals but with diminishing force. Then might not this be the explanation of earthquakes?[24]

Now in writing *Clavis Cœlestis* Wright had studied Whiston's *Astronomical Principles of Religion*,[25] and there he must have read a remarkable passage concerning the interiors of the Sun, planets and comets:

> If there be any such Cavities and Recepticles for living Creatures, and the Things necessary for their Sustenance, in the Central Regions of the Sun, or of the Planets, or Comets, 'tis certain their State and Circumstances must be very different from those on the Surfaces of the Planets. They must all live in Concave Spheres, which must hinder all Intercourse

---

[20] *Original Theory*, p. 43. Cf. *Second Thoughts*, Preface, p. 8.
[21] Letter I, p. 4.
[22] Letter I, p. 5.
[23] Edmond Halley, 'An account of the cause of the change of the variation of the magnetical needle; with an hypothesis of the structure of the internal parts of the Earth', *Philosophical Transactions* 16, no. 195 (1692), 563–78, p. 568.
[24] Letter I, pp. 6–7.
[25] W. Whiston, *Astronomical Principles of Religion, Natural and Reveal'd* (London, $_1$1717, $_2$1725). For evidence that Wright studied this work, see the present writer's Preface to the facsimile of Wright's *Clavis Cœlestis* (London, 1967).

between them and this visible World: Nor can they have any Philosophical Evidence that there is such an External World at all; which is the case of the rest of this Universe, as to us, if we, with all the visible Stars, Comets and Planets, be our selves included in such a Cavity; which is not absolutely impossible to be suppos'd. But then, as to the particular Circumstances of such Creatures, their way of Living, and the Course of Nature and Providence, and Divine Revelation relating to them, I shall not venture here to propose any particular Conjectures about them; only hinting this, that the Power of Gravity from the External Parts being in this Case none at all, as we have elsewhere observ'd, there may be therein such a World as is that we here see, with the like Sun, Planets, and Comets; only that they must be so much less in Quantity and Largeness, as the greater Narrowness of their Cavities requires: Yet still such as the Imagination will not be able to distinguish from our larger visible Universe it self.[26]

This passage, which, since he does not cite it, Wright may have all-but-forgotten, contains the germ of the *Second Thoughts*. For throughout the *Second Thoughts* the sky is a solid spherical shell in which most if not all the stars are set as volcanos, while comets, nebulae, and new and variable stars are associated with eruptions from these volcanoes:[27] a nebula, for example, may be formed of celestial lava, while comets are ejected from volcanos in eruption and either refuel the Sun and other solar (or true) stars, if any such there be, or else fall back again to the sky. In particular, the Milky Way

is looked upon as no other than a vast chain of burning mountains forming a flood of fire surrounding the whole starry regions, and no how different from other luminous spaces [nebulae], but in ye number of stars that compose them, or where there are none, in the vast floods of celestial lava that form it.[28]

According to Letter I of the *Second Thoughts*, astronomical rather than moral issues provided the initial motivation for Wright's departure from his lifelong belief that the visible stars are all other suns, and this is strikingly confirmed by the first section of the Epigenoma, entitled ' On the magnitude of the visible creation ': for here the carelessness with which Wright has pieced together his draft has preserved for us a world view intermediate, both in conception and in the development of Wright's thought, between the picture presented in the *Original Theory* and that presented elsewhere in the *Second Thoughts*. The world view of the Epigenoma is intermediate in conception, for it preserves his earlier picture of a system of stars, the Sun among them, surrounding the ' more excellent body from whence the whole is govern'd ';[29] but

[26] 1st edn, pp. 95–96.
[27] E.g. Preface, p. 9; Letter I, pp. 8–9, 20, 28.
[28] Epigenoma, p. 8. Cf. Letter I, p. 28.
[29] *Ibid.*, p. 4.

it introduces the doctrine that besides these stars properly so-called

> there are infinite others fix't in ye solid firmament of heaven which have in every respect a like appearance, but are in reality no other than vast ignovomous fountains of etherial fire, or inflammable matters, ordaind by the Infinite Wisdom of God, no doubt, to feed the numerous Sun or solar bodies with perpetual fuel.[30]

Further, it is intermediate in time, for in the relevant calculations Wright uses for the distance of the Earth from the Sun the round figure of eighty million miles, as in the *Original Theory*.[31] This figure does, it is true, appear elsewhere in the *Second Thoughts*,[32] but in Letter II he uses the results of observations of the transits of Venus across the Sun (which took place in 1761 and 1769) to derive a value of nearly 91,000,000 miles.[33] Fortunately in each case the calculation is integral to the text, and unless Wright has grossly blundered we can conclude that the relevant section of the Epigenoma antedates that of Letter II. And in Letter II, as generally in the *Second Thoughts*, it is the Sun which is at the centre, and *every* star is a volcano in the solid firmament. For the first time in Wright's thought, then, the Sun is not a star; instead, it has become the unique focus of the visible creation. In his dream[34] of the Fortunate Islands, Wright had made the Temple of the Sun the centre of their worship; now he can urge *us* to turn to the Sun when we meditate on the Creator.[35]

What reasons led Wright to place the Sun at the centre? No certain answer can be given from the patchwork of manuscript that survives, but it may well be that speculations such as those already quoted from Whiston suggested to Wright's fertile imagination that the firmament viewed from without is a sun or planet of a higher-order system, and that the volcanic and fiery nature of the firmament made it more likely to be a sun rather than a planet. But in that case, by analogy, one would expect a luminous sun at the centre of *our* creation, and if *a* sun, why not *the* Sun? Yet, whatever the initial reason, Wright having placed the Sun at the centre betrays two tendencies of thought: when speculating at the physical level, notably in Letter II, he will consider possible variations of systems-within-systems and will even envisage an

---

[30] *Ibid.*, p. 3.
[31] *Original Theory*, p. 71.
[32] Letter I, p. 33; Letter II, p. 4.
[33] Letter II, p. 31.
[34] 'Fortunate Islands, or a discovery of the new world. Designed on a type of all human & earthyly perfections', in twelve letters, Newcastle Wright MSS, Vol. I.
[35] Letter II, p. 36.

infinity of separate systems scattered throughout space, as in Plate XXXI of the *Original Theory*;[36] but when his concern is to reconcile the physical and moral aspects of the creation, as in Letter III, he favours a more precise picture in which the Sun and the firmament are two concentric spheres of a series extending infinitely both inwards towards the supernatural centre and outwards away from it:

> Let us suppose then an other centeral body with in ye orb of the Sun, and in a great degree similar in powers or operations to that without it, i.e. with a like system of planetary bodies in that internal sphere. Let us imagin if you please at ye same time an other still more vast external system of bodies as circumscribing the visible creation, the starry regions, to us being only ye internal concave of their more inormous and immeasurable Sun.[37]

The reward of a virtuous life consists in promotion to a more spacious existence in the sky; the penalty of vice is not crudely-administered suffering, but a more confined and, so to speak, unfulfilled existence in the miniature world of the Sun.[38] But, we might object, if God is especially present in the unique centre of his creation, surely beatitude consists, as it has done since 1734, in proximity to the centre? No, for a man relegated to the Sun is in reality as far as ever from the centre, for the centre is still an infinity of concentric spheres away:

> ... the Heaven of one state or creation, may prove little more than the Hades of an other, and so on ad infinitem both ascending to infinity and descending to negation: magnitude & miniature having no proportion or distinction in ye ideas of God, God himself being only the inapproachable center and circumsphere of his own illimitable nature.[39]

With this vision of 'orb circumscribing orb',[40] Wright has arrived at last at a world view that integrates the physical and the moral order. As an astronomer he has explained variable stars, comets, nebulae and the rest, phenomena that had been perplexing his contemporaries; and he has answered to his own satisfaction the two main arguments against his new account of stars and comets, claiming that the observed motions of individual 'stars' are merely displacements within the craters of the volcano,[41] and that the alleged identification of one of the comets appearing in the late 1750s with that forecast by Halley was, if not mere coincidence, then a periodic eruption just as one would expect of a volcano.[42]

[36] Letter II, pp. 10, 11, 26. Cf. Epigenoma, p. 4.
[37] Letter III, p. 8; cf. *ibid.*, p. 17.
[38] E.g., Letter III, p. 5.
[39] Letter III, p. 6.
[40] Letter III, p. 17.
[41] Letter II, p. 28.
[42] Letter I, p. 15.

As a moralist he has provided the creation with a focal position where the Creator is especially present, and he has supplied an infinite series of spheres concentering on this focus, each sphere a sensorium of the Deity[43] and each a step in a gradual progression from less to greater beatitude, so matching spatially the gradual progression which his contemporaries admire so much in the Chain of Being that stretches from the lowest forms of life through Man to the angels.[44]

Scrappy and diffuse though the manuscript is, the *Second Thoughts* marks the culmination of Wright's life-work. Like the *Original Theory*, it contains passages that are remarkably forward-looking, such as his opinion that undiscovered planets exist beyond Saturn or within the orbit of Mercury,[45] that a planet between Mars and Jupiter has been fragmented by collision with a comet,[46] or that the planets spiral slowly into the Sun over millions of millions of years and so may be described in developmental terms, as reaching the prime or upon the decline.[47] But to extol the work because of these would be as ill-conceived as to condemn it for its solid sky set with volcanic stars: the *Second Thoughts* brings within sight of success the reconciliation of science and religion in spatial terms begun in the 1734 lecture and developed in the *Original Theory*, and it serves to set that much-misunderstood work in the context to which it belongs.

****

In the transcription of Wright's manuscript, the original has been followed except with regard to Wright's use of capital letters and, occasionally, punctuation. The few editorial amendments are placed in square brackets ([,]). The manuscript has been assembled from various drafts composed at different times, and inconsistencies are not infrequent; the more important of these are indicated in the notes. The manuscript appears to be complete except for three pages (Letter II, pp. 19–21) and the plates, of which only one survives. Some of the plates are described in passages integral to the verbal text, others in interpolations added in a convenient space; Wright's 'Index to the plates' has been

[43] Letter III, p. 4.
[44] Letter III, p. 6. Cf. Preface, pp. 3–4.
[45] Letter II, pp. 3, 14.
[46] Letter I, p. 34.
[47] Letter III, p. 31.

extended to include all of these descriptions. Wright's page numbers are indicated by figures enclosed in square brackets in the text. Unless otherwise indicated, Wright's footnotes occur on the page to which they refer.

Grateful acknowledgement is due to the Curators of Durham University Library for permission to publish this work, and to the Librarian and to Dr. A. I. Doyle, Keeper of Rare Books, for many kindnesses.

Second, or Singular Thoughts upon the Theory of the Universe

Designd as a Suppliment
or Sequel to that Hypotheses,
Founded upon New Observation
and Discoveries by the Author
In 3 Letters

Be not Ignorant of any thing Great or Small
Ecclesiasticus 8.15

Sir,

In the 2ᵈ of my former letters I have endeavourd to shew upon what grounds I had presumed to alter my opinion of the general mode or system of yᵉ Creation, and I believe I need not here confess that with some partial[it]y I have there said every thing I could sugest in deffence of this new consti[t]ueted system. In this I shall further & finally consider it most naturally & rationally, and what may probobly be yᵉ morral consequence of such a connected harmonious & providential disposition of its constituent parts.[1]

---

[1] *This introductory epistle is much altered and may originally have served a quite different purpose, since it formerly began: ' In the former letter I have endeavourd to shew upon what grounds I have presumed . . .'.*

PREFACE[2]

The following work having for a considerable time taken up the authors attention, as a kind of duty to the public, he hopes he shall be indulg'd with a canded review of these his late conjectors, which as his last thoughts upon so sublime a subject he confesses he has much at heart. It may therfore be expected he will naturally be found partial to so interesting a theorie. But as of this every reader will be a sufficient judge and duly able to make just & proper allowances therefor, he thinks he shall nead no further apology for what he hath here produced.

But to prepare all such readers as are not scientifically dispos'd and personly qualified to examin and consider the substance of this short treates, the author earnestly recommends the reading of the following *Spectators* viz: N° 519 & 565, that most especially N° 519 and which for fear it should not be in all his readers hands, the most material or significant part of it, he has here taken the liberty to transcribe:[3]

[2] *Note*
The greatest idea we can possibly conceive of a generated progressive, or measurd duration, is to imagin the Sun to be perpetually increasing in magnitud, or the starry firmament to be continually decreasing, till at last they may connect, unite or coincide with each other and make at length one solid globe.

If creation is a production of Nature from an original chaos, its final state or desolution will also be a chaos again, and may be renovated and so regenerated again ad eternitatem.

The active principal of fire is expansion, dilation, and rarefaction, hence ye quan[t]ity of mater in ye Sun, planets and starry firmament tho always ye same to us may be perpetually evaporating so as in process of time to become one united & infinitely blended mass of undistinguish'd particles and consequently compar'd with infinity phisically speaking nothing.

---

[2] *The 'preface' is pieced together from three sections: pp. 1–4, presenting material from* The Spectator; *pp. 5–7, matter which was originally headed* '*Scholia*'; *and pp. 8–15, which were once labelled* '*A*', '*B*', . . . , *and which may form the original Preface.*

[3] *Number 519, by Addison and dated October 25, 1712, deals with the* '*scale of beings*' *below man, and then goes on to speculate about the continuation of the scale above man. Number 565, also by Addison and dated July 9, 1714, treats of the stars as other suns, and continues:* '. . . *when I still enlarged the idea, and supposed another Heaven of suns and worlds rising still above this which we discovered, and these still enlightened by a superior firmament of luminaries, which are planted at so great a distance, that they may appear to the inhabitants of the former as the stars do to us*'. *It also mentions Newton's conception of space as the sensorium of God. Wright has deleted from his ms. numbers 575 and 635, but he has at the foot of the page jotted numbers of issues of* The Spectator *with speculations on the life to come. For another reference to Addison, see* Original Theory, *p. 36.*

[3] If the scale of beings rises by such a regular prograce, so high as man, we may by a parity of reason suppose that it still proceeds gradually through those beings which are of a superior nature to him ; since there is an infinitely greater space and room for different degrees of perfection, between the Supreme Being and man, than between man and the most despicable insect.[4] The consequence of so great a variety of beings which are superior to us, from that variety which is inferior to us, is made by Mr Locke, in a passage which I shall here set down after having premis'd, that not withstanding there is such infinite room between man and his Maker for the creative power to exert itself in, it is impossible that it should ever be fill'd up, since there [4] will be still an infinite gap or distance between the highest created being, and the Power that produced him.

Mr Locke says

That there should be more Species of intelligent creatures above us, than there are of sensible and material beings below us, is probable to me from hence; that in all the visible corporeal world, we se[e] no chasms, or gaps: all quite down from us, the descent is by easy steps, and a continued series of things, that in each remove differ very little one from the other.

and again a little further he says

When we consider the infinite power and wisdom of the Maker, we have reason to think that it is suitable to ye magnificent harmony of the universe, and the great design and infinite goodness of the architect, that the species of creatures should also, and by gentle degrees ascend upward from us, toward his infinite perfection, as we see they gradually descend from us downward.[5]

[5] Perfect knowledge and infalliable truths are not to be expected from human nature, but yet sufficient lights into our existing principles, for the purpose of vertue and morality, is not to be dispard of. And tho' it is impossible for human reason to penetrate beyond the starry firmament and without the limets of the visible creation, yet provided we conjecture nothing unworthy of ye Devine Nature, or that leads us to impiety and infidelity, it is no doubt not only laudible but the indespensible duty of human natures to lay hold of evry circumstance that can inspire us with a

---

[4] *On this common eighteenth-century conception see Arthur O. Lovejoy, The Great Chain of Being (Cambridge, Mass., 1936), pp. 190 seq.*
[5] *John Locke, An Essay Concerning Human Understanding, III, 6, §12.*

love of our Creator, and reveal to us a natural knowledge of his Eternal Being (otherwise faith would be without its object); so far at least as to establish a rational and religious hope of a future state of happiness and imortality. In God alone is perfect truth, and as he is evry where without motion, he knows evry thing without resoning or reflection. But finite natures whose faculties are all confind and limited by sense, are only capable of a ratiocination suited to their present state. [6] Reason, experience and observation, infinite Wisdom seems to have thought sufficient for all the discoveries, necessary to ye human understanding in all its branches of judgment & knowledge towards leading & directing it to the Supreame Good, and Eternal Happiness, and long before any other lights were receiv'd or admitted in the world these *attributes* if I may call them so were deified by ye antients under ye common name of Graiae[6] or Sagasities, they had but one common eye which was ye Sun, and one common tooth, which was time, that preyd upon all Nature.

These were said to have been born grey, and from ye moment of their birth ye naturall nurses of all oracles, information & prediction, long before we had any knowledge or idea of revelation, prophecies or inspiration upon Earth. But of this I shall have to say more in its proper place.[7]

The eternal truths of God require neither miracles or mistery either to confirm or conceal them, since the very lowest or most illiterate of human beings [7] are as capable of instruction as ye most learnd in arts and sciences, and tyrany alone to preserve its own usurp^y athority, seams to have invented many modes, both of devine and human worship for its own all ruling purposes of absolute & sovreyn power.

The powers and operation of human nature are all mechanical and artificial, but those of the Divine Nature are all generative and natural, and we must be very careful to attribute nothing to ye invisible powers of Providence that is repugnant to its visible laws, & of which our senses and our judgments are ye tests. Otherwise we shall render the great God of Nature no other than a Prometheus, Vulcan or Pygmalion and tho this might have been admitted in ye days of darkness, it is the duty [of] our Divine & Holy Religion to rectifie this gloomy errour & reject it now.

[8] Since I publish'd my *Theory of the Univers*, and [being] by no

---

[6] *The Graeae, sisters of the Gorgons, figure in the legend of Perseus.*
[7] *Not, fortunately, in the present work.*

means reconciled to the present theory of comets and many remarkable changes in the stars, a most astonishing idea came into my head, which I have never been able to eradicate, and is now settled into a fix'd opinion: which must in its consequences open so vast a field of speculation, to all curious naturalist[s] & particularly all astronomers, that I can't help thinking it my indespensible duty to communicate it to the publick.

With human beings matter space & time are all of a very finite nature, but in ye head or hand of om[n]ipotence, they have no dementions, and hence all magnitude and miniture to ye Eye of Providence are as equal objects. In this one great attribute of ye Devine Nature, is comprehen[d]ed all the mistery of creation, which to mortal natures & created beings must be ever incomprehensible.

[9] The many and various luminous spaces in ye celestial regions which accompany the stars, and bright[ness] mid them are look'd upon by all modern astronomers to be no other than an indefinite number of small stars, not to be distinguished by ye best telescopes.[8] But by this hypotheses in which the stars are universally conciv'd to be so many vast vulcanos, are next to demonstration clearly prov'd to be vast floods of burning lava, such as frequently issue from ye erruptions of Etna & Visuvious, & all other vulcanos upon this Earth: the difference only (if any) arising from the unknown nature of ye celestial matters of which they are compos'd.

This is now become so evedent from their general phenomena, that few words, are necessary to convince every rational astronomer of ye real truth of it.

In the account of all new, cloudy and now extinguishd stars &c which have from time to time been observ'd by ye most emenent astronomers, and hitherto, no how satisfactorily explained, by any author tho attempted by many. [10] To those who can arive at any tolerable conception at the great and natural construction of ye vast fabric of ye visible creation, must of force acknowledge it to be ye incomprehensible production of infinite Wisdom and Power and consequently from ye most rational convixtion, be brought to a due sense of there dependence upon that Providence which produces it and to the most awful adoration of his Devine Nature. Thus the devine astronomer *Aratus* drew all created

---

[8] *The view of most astronomers, but not all: cf. 'An account of several nebulae or lucid spots like clouds'*, Philosophical Transactions 29 (*1715–16*), pp.. *390–2*

Since I Publish'd my Theory of the Universe, and I by no means Reconciled to the Theory of Comets and many Remarkable Changes in the Stars. A most astonishing Idea came into my Head, which I have Never been able to Eradicate, and is now settled into a fixed opinion: which must in its Consequences open so vast a field of Speculation, to all Curious Naturalists & particularly to all Astronomers, that I can't help thinking it my indispensable Duty to Communicate it to the Publick.

With Human Beings Matter Space & Time are all of a very finite Nature, but in ye Head or hand of Omnipotence, they have no Dimensions, and to all Magnitude & Minuteness, to ye Eye of Providence are as Equal objects. In this one Great attribute of ye Divine Nature, is Comprehended all the Mistery of Creation. which to Mortal Natures & Created Beings, must be ever Incomprehensible.

Preface, p. 8. Wright hints at the genesis of his new cosmology.

I shall not descend so low as to account for y.e same operations in Bombs and Water-rockets, but can't help observing, that solid heated bodys of any Combustible Matter is undeniably capable of such effects & operations.

To conclude all the New, and vanishing Stars, together with the Clouds and other variable stars that are frequently perceived in y.e Celestial Regions, will by this Hypothesis be naturally and almost incontestibly accounted for, and without any of those forced conjectures which astronomers have been reduced to in attempting to explain them.

The first will be found to be no other than fresh Eruptions, or y.e Revival of an old Volcano; the second such inflamatory substances totally taking fire; the Cloudy such as are more prodigiously circumambient by vapours and the variable, with all periodical ones, repeated Eruptions of y.e same Star at different intervalls & sometimes new at y.e same.

The via Lactea and all the Luminous spaces will likewise come under the same dispositions, or being subject to y.e same Laws & politicks.

Letter I, p. 28. Wright summarises his new explanation of celestial phenomena, including the Milky Way.

beings into ye existence of one omnipotent *God*, and even St Paul[9] has given his testimony to this great truth, *Acts* 17th: 18. &[c]. We nead not therefore be ashamd of such enquires when we find Cicero, Cesar, & Germanicus,[10] in the height of their greatness & power emploid in the translation of this their much & probably their most esteemed author. [11] If any thing can sow the seeds of vertue erradicably in the human breast, it must be a true knowledge of our selves, and dependence upon the infinite Being. I know nothing more likely to do it, than such philosophical enquires as tend to a clear understanding of both.

Knowledge will always have ignorance to struggle with; but the fetters of reason has long been struck off, and the human understanding emansipated, so far as to be fread from any danger in oposing reciv'd opinions, which anciently was so very great as to hoodwink the human mind and shut out every ray of Devine Knowledge. We are told[11] that Anaxagoras was accus'd of impiety, and would have suffer'd death for having affirmd the Sun to be a burning mass of fire, had not his eloquent pupil *Pericles* defended his approv'd master, so as to soften ye sentence into a fine and banishment.

Things are only extraordinary from having nothing of the kind to compare them with. But if we had opportunity & capacities to observe more of ye great oporation of Nature with adequate abilities, to compare one part with [12] an other; for the whole we must never expect to comprehend. Then perhaps without much presumption we might determine somthing nearer the truth than our present ideas have reach^d. [13] The darkest ignorance, and the blindest prejudice, can only be suppos'd, capable of objecting to what is here founded both upon observation and reason, or endeavour to observe any of the features of so noble and interesting a scene. But as the brightest flame cannot aspire, till it has drove away the smoke, a work of this kind must not be smouther'd, and ye fire within conceild for want [of] a little air, and possibly the more vent we give it, the rising blaze may burn the clearer. Ignorance whose attributes are always amazement & astonishment,

[12 cont.] Empedicles, Anaximenes & Diogenes all belev'd the stars to be ignivomus.[12]

---

[9] *St Paul's speech at Athens on 'the unknown God', in which he quotes Aratus.*
[10] *Rather, Cicero and Germanicus Caesar.*
[11] *In Plutarch's* Life of Pericles.
[12] *Ignivomous = vomiting fire.*

require to feed with somthing that looks like miracle, for every new idea to ignorance is equally inconceivable and surprising; but solid sense reserves it[s] wonder & restrains its admiration within the bounds of reason, and seeks no other satisfaction than [14] reconciling what it sees or hears to ye laws of probability, especially where trouth is much desir'd, and certainty can not be had, & hardly to be expected.

I can not more properly conclude this Preface than with ye following lines out of a poem on ye Creation by Mr Baker:

[15]   Each seed includes a plant; that plant, again,
Has other seeds, which other plants contain;
Those other plants, have all their seeds, & those,
More plants, again, successively, inclose.
5   Thus every single being that we find,
Has really, in its self whole forests of its kind.
Empire and wealth, one acorn may dispense,
By fleets to sail, a thousand ages hence;
Each myrtle seed includes a thousand groves,
10   Where future bards, may warble forth their loves.
So Adam's loins containt his large posterity,
All people that have been, & all that e'er shall be.[13]
Amazing thought! what mortal can conceive,
Such wondrous smalness! Yet we must beleive
What reason tells: for reason piercing eye
Descerns those truth our senses can't descrye.

Henry Baker[14]

---

[13] On 'preformation' theories in embryology, see J. Needham, A History of Embryology (2nd ed., Cambridge, 1959), espec. pp. 205 seq.

[14] Henry Baker, F.R.S. (1698–1774) published in 1727 The Universe: a poem intended to restrain the pride of Man *from which these lines are taken* (1808 ed. (Taunton, England), pp. 38–39). *Wright was keenly concerned for poetry, as his manuscripts show.*

LETTER Y^E I OR 10[15]

El sabio muda consejo, *el necio no*.[16]

Sir,

The love of truth and a natural desire of all possible rectitude having induced me to revise my *Theory of the Universe*; I once more must desire you to listen to me a little upon that subject, having great reason from some late researches and discoveries to suspect the bases upon which that hypotheses was plan^d, and notwithstanding the many established athorities of the most eminent of the antient as well as modern philosophers produc'd in its introduction to support it, upon the whole it may possible be found not altogether free from errors, and in some degree perhaps a little doubtful.[17]

This probably may appear to you to be a very unusual and perhaps a prejuditial kind of recantation, capable of prepossessions much to the author['s] disadvantage. But as the Temple of Truth to him, is a much more attracting edifice than that of Fals Fame; tho he is in some measure constraind and necessitated in this short work to sacrifice some of the labours, and plausable advances of many more learn'd & sagatious men who have past their full probation thro' the sciences with applause and particularly upon this subject with public sucess. Let it [2] be candidly observ'd by you at least that in this general and unavoidable sort of proscription of *authors* he has neither spar'd him self or favour'd his own offspring.

What the severale astronomers referd to in my former work have said in the deffence of their opinions; which I must own as intirely agreeable to my own reasoning and conceptions I had so much faith in as to adopt & follow; may be seen in ye several citations of y^e (I) Letter and it may also in ye same Epistle be observ'd that the doubtfull foundation upon which I there trod, was little more than a mear pilework of opinions fram'd and connected together only by my own credulity & partiality to the illustrious authors. But here I hope confiding in my own sole original ideas

---

[15] *Again, as the renumbering shows, what follows has been pieced together from various drafts and extensively revised. The* Original Theory *consists of nine letters, so that this first letter of its sequel could properly be numbered tenth.*

[16] '*The wise alter their counsel, the foolish do not*' (*Spanish*).

[17] *A previous version of this much-altered sentence concluded with* '*fallacious*'.

and more mature judgment you will as candidly allow I have advanced no implausable rational[e], towards removing my natural prejudices or rather acquird sentiments, to a more substantial & solid ground.

You are the only person at present indeed that I propose shall reap any benefet from these my new connected ideas, if any advantage can be said to flow from them? But if I am fortunate enough to make you so far master of this [3] new system of creation as to be willing to beleive with my self, that I have some little reason if not demonstration on my side, as far as I have gon, it will be a sufficient satisfaction to me, to hope that what I have here advanc'd may possibly live to make its appearance in the world, and not as Mr Dryden has some where said, *Die in thinking*.

Now if you please I will proceed to the business in hand. The expected return of ye comet[18] of 1682 in the year 1758 was to have finally prov'd and confirmd to a demonstration Sr Iasac Newton lear[ne]d and ellaborate theory of all those most surprising & erratic bodies in generall, and especially as modern observation could only be depended upon. But as no regular return of any such comet has ever been fully provd or pretended, it follows that the presumptive theory of that great & sagatious mathematic[ian] still remains doubtfull, and would of itself be sufficient to set astronomers again at work to find out new Laws of Motion to solve their astonishing & almost irreconsilable phenomenae. [4] But I now begin to be convinced, that things of this sort are never found out as the effect of labour, or intense study, nor indeed any other intirely new discoveries; all originall ideas generally arising from some fortuitous impulse, or observation founded upon some providential accidents, (if I may be allow'd ye expression) or great operation of Nature which before was either insufficiently attended to or perhaps overlookd. To this great end, time as I have represented before,[19] the great parent of all interesting events never fails to assist us, unveiling with all his powers & in due periods of motion and particularly by some striking circulating & allarming operation of nature, to lead evry attentive mind as by degree to all ye truth and certainty which our present state requires.

---

[18] *Halley's comet. On each return the comet is affected to a greater or less extent by the gravitational attraction of the planets near which it happens to pass, and this causes the intervals between successive returns to vary substantially. Clairaut, but unknown to Wright, had already taken these factors into account and had predicted the date of the comet's return to within a month (A. C. Clairaut,* Théorie du mouvement des comètes *(Paris, 1760)*).

[19] Original Theory, *p. 12.*

And I must ingenuous confess, that neither the failing in Sir Iasac Newtons theory of ye comet, or much less my own yet unconfirmd hypothesis of ye stars, and visible universe, had ever been molested by me but upon accidentaly thinking out, the following lights, & which I can't help thinking it my indesp[ens]ible duty to communicate to my learn'd friends.[20]

[5] The repeated and extensive shocks of the late earthquake at Lisbon,[21] lead me to suspect some other latent cause more probable than the ordinary one generally advanced towards the solution of that allarming effect and unacounted oporation of nature. The first idea that struck me imagination as a possible and rational one, was the now established and admitted formation of ye Earth with a suppos'd abyss of waters or other matter in the center of it, as inclosd within the terrene crust or mundain shell & which may be well concev'd to be of no great deapth or considerable thickness, with regard to ye internal and more expanded submundain space*. In consequence of this not unnatural conjecture I was led to apprehend, with Dr Hally, that there might also be probably a centeral ball or body of a more solid matter than ye outward crust very different from any that has yet been discoverd, upon, or near the surface of the Earth. This Terella or invisible body of matter, that very sagatious astronomer conjectord to be a magnet globe, and in a spherical mañer revolving upon an axis like the similar and external world.**

[6] From this not implausible hypothesis it is that he indeavers to solve the perpetual variation of the mariners compass, and all ye other vague & fluctuating affection of ye no[r]th nedle & magnet. And I can't help being of opinion further that a new and more satisfactory theory of the tides, as well as of all ye known currents, &c, & much less exceptionable than the present Newtonian one, may likewise possibly be produced from it. But upon this topick I hope to have an opportunity of saying much more in its proper

---

* So Sr Isaac Newton, Dr Burnet & Mr Whiston, upon this subject.[22]
** See his *Theory or Solution of ye Magneticall Variation*.[23]

[20] *There is here a break in the manuscript, and the next page has evidently been drafted independently.*
[21] *That of 1755.*
[22] *Cf. Newton,* Principia, *Book III, Prop. X; Thomas Burnet,* Telluris Theoria Sacra, *Libri duo priores (London, 1681), espec. Book 1, Ch. 5; William Whiston,* A New Theory of the Earth *(London, 1696).*
[23] *Edmond Halley, 'An account of the cause of the change of the variation of the magnetical needle; with an hypothesis of the structure of the internal parts of the Earth'* (Philosophical Transactions 16, no. 195 (1692), pp. 563–578).

place, having collected material and gone a considerable way towards framing a new theory of ye Earth,[24] upon those suggestons.

Thus far premisd, I began to think that as the said internal abyss of waters might possibly partake of some piculiar mode of motion, they must of course in time wear or insinuate them selves into ye internal coat or concave shell of ye Earth, and by that means at length loosen or lessen its partiall adhesion or connection to ye generall mass, so as not being able to support it self in its natural bed or concentric arches against the more powerfull centripetal force or general attraction or of some other more powerful influence[25] it would inevitably fall or be projected towards ye center[1] body. For it was natural to conclude that unless it was specifically lighter than ye inferior water it could not possibly float within the great abyss.

[7] This any one conversant with natural and experimen[tal] philosophy must allow could not possibly fail of being attended with all the visible and sensible effects attributed to ye natural consequence & phenomenon of earthquakes and will likewise account for all or most of the many repeated and less virulent subsequent shocks as communicated all round ye central effect & as all derrivd from ye same original cause, for as naturaly following the falling mass, in its prone rush towards the centerl body, the vast and perturbated abyss of water would imediatly & with equall violence and rapidity indeavour to restore it self to an equilibrium. Such repercussion would almost be as effectual as the first disjunction, but yet as naturally to be expected less & less, in each repeated shock till all the central waters were again subsided and the great abyss restord to it quiescent state or pristin tranquility. But this particular subject at present I shall leave to be discust by other philosophers, and proceed to the particulur objects it naturally directed, and drew my attention to, as giving the first conception to a subject which I judged to be of much more importance in the fields of science, and evry way worthy of the most mature consideration.

[8] Upon thus conciving ye idea of a centeral globe and a circumambient sphere of waters floating round it, I venterd one step farther[26] and was willing to imagine, that in larger orbs, and more

---

[24] *In Wright's own list of his manuscripts (Newcastle Public Library, Wright MSS, vol. iii) there appears under ' Very valuable works ' the title ' Theory of the Earth '.*

[25] *A deleted footnote referred to ' explosions, or internal erruptions '.*

[26] *As had William Whiston in his* Astronomical Principles of Religion *(London, 1717), pp. 93–96, a work which Wright used extensively in writing* Clavis Cœlestis.

immense globes of matter, there might possibly be also vast regions of air or aether with centeral spheres of fire and other bodys included. Likewise within them more indefinite shells also of ye same or not unsimelar construction planetary systems might exist and subject to like laws of oporation, and motion with our own world.

Hear then give me leave to look up again to ye stars; for this idea immediatly presented itself to my redy imagination that a new and very possible modification of the visible universe might be derived from it, beleiving it not improbably that the visible heavens or stary firmament might prove to be no other than a solled orb[27] of this stupendious nature and the fixd stars no more than perpetual lumination or vast erruptions & of refulgent or inflammable mater promiscuously distributed as celestial vulcanos allround the starry regions emiting an etherial & intense fire of various magnituds, but remov'd for some infinitely wise purpose of the Creator, to so indefinite a distance as to be far with out ye reach of human arts to ascertain but by ye mental eye of reason only.

[9] This idea which naturally solves from like visible causes and effects, all the several phenomena of the celestial regions without excepting any one of them I am very much inclined to conclude will be found in the end to give the truest construction of the visible creation in all its movements modes, and consequences, of any hypotheses hither to advanc'd.

The comets with great or rather inconcevable force being ejected from those vast erruptive vulcani, in all directions and some in hyperbolic trajectorys, and returning again by their natural gravity or attraction, to the celestial & solid siderial concave, solve in the most obvious maner all the several appearances of those astronishing bodies. The new stars being also of ye same nature and production, ejected in like maner out of their firey fountains but with less force & in a more perpendicular direction are accordingly found to vary only their respective magnitude & sintillation. Those tinged of various colours are all as evedently accounted for by the same principles of nature, and such stars also as are now extinct, are thus evinced and prov'd to have been no other than like vulcani having long since seas'd to emit their flame and burning matter. [10] The incredible vicinity of the comet of 1680 to the center of ye system, which according to Sr Isaac Newtons theory of the motion of these erratic bodies must have approach'd within less than a sixth part of ye Suns diameter[28] is so irrecon-

---

[27] *Whiston*, ibid., *p. 96*.
[28] *Newton*, Principia (*2nd ed., London, 1713*), *p. 480*.

cileable to reason, not withstanding what that great philosopher has said in support of it, that I am perswaded had he liv'd to have revis'd that learnd and ellaborate work once more & by ye help of more accurate observation, to a person of his great abilities, and penetration, many new doubts and difficulties must have occur^d that unavoidable would have induced him to have alter'd his opinion, and in consequence of which to have fram'd other conjectures more rational both of their origin, and laws of motion.

One result and very natural consequence of such vast excentric orbits is that in evry revolution of most of their immeasurable trajectors they must pass thro such intollerable changes of the most intense heat and cold, that no possible subst[ance] or matter of a perishable nature, can be supposd to exist in it. [11] That of 1680 in particular Sr Isaac tells us in his *Principia*,[29] near its perehelion point acquir'd heat 2000 more inf[l]amatory than red-hot iron.* Add to this that as their tails are always observ^d to fall nearly in a direction opposite to ye Sun, in passing the perehelion points of their orbits this constantly observed direction must always be preservd in it visibil[e] & unalterable possition by an almost insolvable velocity.

But to this I am well aware that a mathematical casuist may plausably answer that the approximating tail, may possibly be totally alter'd or dissipated in the Suns beams, and a new one probaly generated, as it receeds from from the Sun, but of this I must own I shall not easily be convin[c]ed. Besides the formentioned comet during its short stay with in y^e sphere or orbit of Mercury, which could only be from the 5th to y^e 12th of December, it must have mov'd with such a velocity that the extremity of it tail could not describe an arch or curve less in extent than y^e whole orbit of that planet, and in all probability considerably more, and even $\frac{2}{3}$ of it in one day. Hence I may conclude its least velocity to have been equall to near 100,000 miles per minute, and y^e greatest almost 4 times that space. [12] These consid[er]ations are abundantly sufficient to have calld any athority, but Sr Isaac Newtons in to question. But as revering his allow'd abilities & establis'd applause I never suffer'd my self to entertain a dout of any of the elements he has so learnedly advanced.

[12] * The same comet in its aphelion, from like principals of demonstration may be prov'd to have past thro a degree of cold 4000 times more intence than freezing ice.

---

[29] *Newton*, Principia (*1st ed., London, 1687*), *p. 499.*

But observation the great progenitor of all discoveries has at length furnish^d us with materials far more evincial than reson only of its self, could reach. And[30] first *to do justice to ye great athor mentiond* above I shall beg leave to add:

[13] The elements of three distinct comets suppos'd by Sr Isaac Newton to be only the periodical returns of ye same and upon which was founded the expectation of a like appearance again of ye same comet in ye year 1758:

Anno Dom	M[onth] D : H : M	Perehelion ° ′ ″	Dis: a ☉	Nod ; asc. [° ′ ″]	Inclination of ye orbit [° ′]
1531	Aug^s: 24 : 2[1] : 18	♒ : 39 : 0	56700	8 .19.25. 0	17 : 56 Retrograd
1607	Octb: 16 : 3 : 50	♒ : 16 : 0	58680	8 .20.21. 0	17 : 2 Retrograd
1682	Sepm: 4 : 7 : 39	♒ : 52 : 45	58328	8 .21.16.30	17 : 56 Retrograd

From these elements construct by Dr Hally,[31] it evedintly appears how naturall it was to have expected somthing like a return of ye same if a regular body some time within ye year 1758. But at ye same time we may also gather very sensibly the improbability and uncertainty of its appearing to any fixt moment.

The first period or time betwixt ye first & second perihelion being 76 years & 52 days; and that of the second revolution taking up little more than 74 years & 323 days making a difference of 1 year & 94 days in point of time; a sufficient *prostavering*[32] to show that we could have no soled grounds of conjector to ascertain the precise time of any such phenomenon, to which if we add the effects of a large menstral, if not diurnal parallax which will naturally attend all such trajectories, we must confess our selves as much at loss to know [14] in what part of y^e heavens to look for them, and likewise the little dependence we can possible have of even y^e most correct observations.

If therefore the difference of the 3 theories arise only from y^e inacracy of observations, the two per[i]ods may very rationally be reduced to a mean of about $75\frac{1}{2}$ years; it was proba[b]le therfore to have expected some appearance of this commet towards the latter end of that year or the beginning of 1759.

Providence no doubt has a very watchfull eye of all the divine ordinations, and hence we may conclude the Earth to be in no

---

[30] *This sentence is an interpolation, allowing Wright to incorporate a separate manuscript on the elements of comets.*

[31] *Edmond Halley, 'Astronomiae cometicae synopsis'* (Philosophical Transactions 24 (1704–05), pp. 1882–99), p. 1886.

[32] '*prostavering*': perhaps for '*prosthaphaeresis*', also spelled '*prostapheresis*', a 'correction' used in planetary theory. The word was also used in a more general sense (see O.E.D.).

danger from any such erratic bodies.³³ But if the motions of comets could be reduced to certainty & as solid and regualar bodies the end of ye Earth might be as certainly predicted. For a point of time of course must be found in which some comet in its nodes might be so near ye Earth as to be attended with very alarming if not fatal consequences.

[15] I conclude therefore that as there is nothing infalibly certain, but the will of God, that those alarming body^s are only & merely accidentle, as all the other great elementary operations of nature are, thô not without being subject to some infinitely wise purpose of the Divine Nature which we can not discover.

It follows therefore that thô the respective theores hear consider'd do not so exactly agree with one another in evry circumstance, so as to prove them incontestibly to arise from one & ye same body yet from them we may draw a sufficient demonstration to prove that as appearing in ye same region of the heavens they may all be ejected from ye same or some near neghboring stars in different states & periods of erruption and even some of them not very far from regular returns or equal intervals of ejection.

[16] To obviate any difficulty in comprehending how such an amazing number of astonishing bodies can be projected [from] ye starry firmament to such an indifinite distance. Tho very probably such an idea never before enter^d into a human breast, let it be considerd that many burning mountains upon Earth and which bear no kind of proportion to ye solid concave region of ye stars, frequently throws out vast masses of burning matter, to an almost incredible distance, in ye same maner, particularly Mount Vesuvius which in the year 1631 threw out vast masses of matter to a prodigious hight, and particularly one in form of a bomb as far as Nola; in a range of about 12 miles from ye mountain; and which set ye Marquis of Lavros house on fire.

But those who are willing to know, to what a prodigious hight burning mountains are capable of ejecting not only ashes, but other more solid & flaming matter I refer to Sr Willm Hamiltons very learn^d account of volcanos, & particularly of Vesuvius & Etna.³⁴

---

[33] *But see Whiston*, A New Theory of the Earth, *Book IV, Ch. 5.*

[34] *Several accounts of volcanos from the pen of Sir William Hamilton were published in* Philosophical Transactions *(57 (1767), pp. 192–200; 58 (1768), pp. 1–14 (with dramatic sketches of the 1767 eruption of Vesuvius); 59 (1769), pp. 18–22; 60 (1770), pp. 1–19; 68 (1778), pp. 1–6; 70 (1780), pp. 42–84).*

[17] All that can be said in support of the present theory of the comets as advanced by that philosopher and funded upon his *Principia* is that in the year 1757 a* comet did appear, and pretty near to yᵉ time it was fortold, but most unexpectedly in a contrary direction and in a different inclination to that of its antecedent in yᵉ year 1682 &c; that soon after also two other comets appeared, one in yᵉ year** 1759 and the other in the following,*** but neither of their elements having yet been produc'd, all that we know of them are a few observations in yᵉ *Philosop[h]ical Transactions* of those years, and one of these has been imagind to be the comet that was from time before expected.

But as no year is suppos'd to pass over without some of these surprising phænomena, thô so few of them have fell under our observation at yᵉ Earth; many in all probability escaping with out a possibility of being seen: no certain inference can be drawn from them as any rational conclusion upon the subject in question; and consequently [18] a proof positive must always be wanting till more correct observation, and a perfect theory of these celestial bodies puts it out of all dispute, but this their indefinite numbers

---

[18] * This comet was observ'd at Gree[n]witch in yᵉ months Sep. and October 1757 and was found to be direct.
  The perehelion Octoᵇ: 21ᴰ : 7ʰ : 55' P.M. : M[ean] T[ime] at G[reenwich]
    Inclination of yᵉ orbit 12° : 50' : 20″, its node in ♉ : 4° : 12' : 50
    Place of yᵉ perihelion in ♌ 2° : 58' : Perʰ a Node Descᵉ : 88° : 45' : 10″
    Logarithᵐ of yᵉ perehelion disᵗ : 9.528328
    - - - - - - - of yᵉ diurnal mot[ion]: 0.667638 [35]
  ** In May 1759. Observ'd by Dʳ Beves, and Mr Munckly and others.[36]
  *** In 1760 by Mʳ Short and the Revᵈ Mʳ Michel and others.[37]

---

[35] *James Bradley*, ' *Observations upon the comet that appeared in the months of September and October 1757, made at the Royal Observatory* ' (Philosophical Transactions 50 (1757–58), *pp*. 408–415).

[36] *J. Bevis*, '*An account of the comet seen in May 1759* ', and *N. Munckley*, '*An account of the same comet* ' (Philosophical Transactions 51 (1759–60), *pp*. 93–96).

[37] *J. Short*, ' *Observations of the comet seen in January 1760* '; *J. Michell*, ' *Observations on the same comet* '; *N. Munckley*, ' *An account of the same comet*'; and *M. Day*, '*An account of the said comet* ' (Philosophical Transactions 51 (1759–60), *pp*. 465–469).

C

and very various directions rather leave no room, or even a shadow of ground to hope for.*

[20] The flux and reflux of this celestial matter, what ever its composition may be, is farther corroborated if not clearly demonstrated by the periodical exhibitions of several new stars, besids a perpetual increase of the brillance of many & as visible a decay in ye light of others, in somuch that some denominated of the second magnitude in Hiparchus & Ptolomys time are now distinguish^d as the first.[40] Particularly Procyon in ye Lesser Dog; one in the Great Bear & another in Orion. Others have also as much

[19] * But the clearest ellucidation of these celestial vulcanos or siderial erruptions will appear in a discovery made by Hugens[38] in ye year 1656. See that great athors works & also Dr Longs *Astronomy* B[ook] 2 Chap. 22. P. 353. § 942, or take it in his own words as follows:

Looking accidentally in the year 1656 thrô a telescope at ye middle star in Orions Sword, I saw twelve stars instead of one, which was no new thing; there position was such as is shewn in figure annex'd: seven of these stars, of which three are very close together, seemed to shine through a cloud, so that a space round them of the shape represented in the figure appeared much brighter than any other part of the heaven; which being very serene and very black, look'd here as if there were an opening through which one had a prospect of a brighter region: I have since often [vi]ew'd this wonderful appearance, which continues the same without any change of place or shape to this time.*

See Plate I

Several other such luminous space[s] as caverns or casms have been observd by other astronomers & full of stars, but not so erregular or conspicu[ous]ly defind. From all of which being evedently sensible to sight, it consequently follows that the sollidity of the starry regions as teraignuous orb can be no longer a doub[t].

[19′] * When Hygins first discover'd this most astonish[in]g phenomenon, about a centry ago; it was then much more conspicuusly defind than at present, proba[b]ly from its being more distinguishable near ye time of its first erruption, since Dr Long from his own observation of it gives us the following and somthing different account:

Hugens who first discoverd this wonderful appearance as he justly calls it has given us a draught of it, but the stars are all drawn nearly of equall magnitude, and the luminous space is more defind than it aught to be,** which faults were in all likelyhood owing to ye mistakes of ye engraver. I have therefore given an other* scheme of it such as I have often seen it thrô a telescope of 17 feet and have exprest there in, the apparent magnitudes of ye several stars.[39]

** Rather & more probably the stars may have varied their appearance as also the lava cooling must also seam less bright.

* Fig 96 Longs *Astronomy*.

---

[38] *Huygens*, Systema Saturnium (*Hagae, 1659*), *p. 8, cited by Roger Long in* Astronomy 1 (*Cambridge 1742*), *p. 353.*

[39] Astronomy 1, *p. 323.*

[40] *Cf. Long*, Astronomy 1, *Book 2, Ch. 22. An early list is that of Halley*, '*A short history of the several new stars that have appeared within these 150 years*', Philosophical Transactions 29 (*1714-16*), *pp. 354-6.*

apparently deminishd and from being originally accounted of the third magnitude are now reconnd only of ye 4th or 5th as that of Merope in the Pleiades &c. But the ebbing and flowing of this siderial flame is most manifest in the ignivomous revolution of the stars following, namely

That of ye Whales Neck in 11 months (fere)[41]
- - - - - - - Swans Breast in 14 years
- - - - - - - Swans Beak in 10 months

all of which alternately have been extinguished & again revivd. Various have been the conjectors towards a rational solution of this astonishing phenomena, but hither to far from satisfactory.

But [21] to illustrate in some measure the previous trajectories in which comets or new planets[42] may possibly appear, or have been observd to approach or receed within the region of the known planetary system, or even in y^e neighbour hood of the Sun, let us

PLATE II

imagin the principale circle[43] in Plate 2 to represent the sphere of vision without the solar system so far as any interfearing object can fall within our observation, and then will the right line $AB*$ represent the projectile of a comet in a true perpendicular direction towards the center of y^e Sun, but approaching no nearer than the most remote planets, or perhaps not so far as to enter any of the known orbits.

In such a case the comet in its first apparent approximation must of course appear without a tail as in a state [of] occultation and increasing magnitude only for a short time, it must soon again deminish in a like proportion and near ye same point of ye heavens, in a few days totally disappear. But if such a body as this should so arive within the power of any temporal or local parallax it would before it totally vanish^d seame to discribe to us some part of a small curve of ye spiral kind. [22] Such a one is represented at E and which having escaped the attention of astronomers in its approach appeared anno 1604 in the constellation Serpentarious, and there only changed it[s] apparent place about 3 degrees, and daily deminishing in magnitude in about ten months time, it totally disappeared. This comet was by Keplar at Prague mistaken for a new star and the comet of 1580 is conjecturd to have been one of ye like sort. Such a one also was observd in December 1315 moving nearly concentric to ye North Pole.

[41] '*fere*', i.e. '*roughly*'. Long (Astronomy, *pp.* 343 seq.) discusses Mira Ceti *and other variable stars.*
[42] '*or new planets*': *this phrase has been inserted.*
[43] *Details of the Plate, though not its number, were in the original draft.*

Secondly if any comet moving in a right line should, as at C, D, O have force or momentum enough to reach the body of ye Sun, or even his inflam'd atmosphere it is very natural to suppose that it would of course be totally dissipated or absorbd in the solar beams or other wise fall into ye Sun. And such may probably be the original of some of[44] the solar macula or spots visible even to a naked eye in foggy weather floating continually [and] perpetually connecting or separating as upon his disk.

These likewise, in more minute particles, may also by a centrifugal or repelling power be evry way disperst in ye direction of his axeler rotation to a much greater distance from him and probably occasion that [23] beautiful Aura Etheria[45] frequently seen in such direction & already known to attend his globe,— particularly in all solar* eclipses.

The comet of 1472 as appears from its mode of motion, and vicinity to ye Earth, as also that of 1450 if not mis-represented by ye observers, may probably have been of this sort. Thirdly such comets as deviate in the[i]r motion by little from a right line, and approach no nearer than the lesser or inferior planets as at I may not only be expected to recead again, but are also sometimes seen with great velocity retiring and atten[d]ed with enormous tails. Such was that of 1680 marked F which has been imagind nearly to have followd the same tract with one in 1006 &c and consequently to have been the same body. But 4^ly those who fall without ye orbis magnus as at G and H moving in less excentric curves and even some others that have fallen within it in wider ranges of ejection—as at K & L—seldom appear but once, the first generall occup[y]ing but a very small portion of the stary hemesphere, but the last a much more extensive tract. Both these if duly observ'd from first to last of their appearance magnifie and diminish in like proportion except where the position of ye Earth in its orbit produce a sort [of] tangent to their trajectories. Of these kind [24] were those of ye year 1556 and 1607, the former moving when swiftest about 15 degrees, and the latter nearly 30° per diem. 5th: all such comets as are at first discoverd, and again

---

* In that of 1715 it was said to be very manifest but not so bright as to obscure the stars, but these may also arise from an other cause, see ye apendix.[46]

[44] 'some of': originally 'all'.
[45] 'Aura Etheria': cf. Wright, Clavis Cœlestis, p. 14: 'The Sun's circumambient Atmosphere, on all sides together, form a vast inflamed Region, called, or rather form[ing] that Medium from whence proceeds, the Aura Ætherea ...'.
[46] No appendix to the Second Thoughts survives.

are seen to vanish in ye same sign of the Zodiac or at most without departing far from the same constellation of stars, are generall subject to ye most irregular motion by reason of their discribing a trajectory very much deviating from that of a great cir[c]le, and sometimes hardly reconcilable to any mathematical solution at all. Sixthly all such as make their appearances and exits only near an opposition to ye Sun can by no laws of optics or of motion be supposed to have either circumscrib^d the center of ye system or to be any how affected by the planetary principals and this which is next to a demonstration is an evedent contradiction of their having any regular or hipothetical connection with our solar body. Se the geocentrical places of those comets in the years—71, 380, 1240, 1315, 1340, 1457, 1473, 1556, 1652, all of which appearing in or near an opposition to ye Sun could not possibl[y] rea[c]h any of ye lesser planets.

[25] The comets cary vast trains of light behind them; the planets vast tracts of darkness.[47]

The planets all in their annual[?] or heliocentric motions universally more direct.

The comets are not confind to this law in all their visible trajectories, many of them moving retrograde [which] motion is quite reverse. The planets are all visibly dependents upon ye Sun, but very few of ye comets are observ'd to reach the orbis magnus.

Thus as in all their principal predicaments they appear to be retrogressive bodies we may venter to doub[t] at least, if not conclude, that they have no common connection with our sistem so as to be lookd upon as regular parts of it but as all of them totally independent & subject to other laws.

All[48] of the above reasons are not equally strong but some of them are unanswerable and thô none of the comets have yet been observ'd to agree in all respects with their several theories, as this might consequently arise from the inaccracy of former observation, it is not of it self sufficient to explode their present doctrine. But every other circumstance & argument impartially considerd, the bases of it at least must sh[r]in[k] a litt[l]e, if it foundation dos not shake.

[47] *Wright here jots down arguments against the Newtonian doctrine that the 'comets are a sort of planets revolved in very eccentric orbits about the Sun' (Newton, Principia (London, 1687), p. 508; Motte-Cajori translation (Berkeley Calif., 1934), p. 532).*

[48] *The following two sentences are an afterthought.*

[26] Hence as the planets are all incontestably govern'd by the same common and universal laws of motion and as revolving all in ye like direction both in their annual and diurnal courses and nearly also in ye same plain are accordingly demonstrated to be regular bodies and subject to ye solar system. But ye comets are found to move in all mañer of directions and consequently many of them directly contrary to each other, were their no other reasons to suspect their being regualar bodies and constituent parts of our visible creation this would of its self be sufficient to call any known theory of these surprising bodies in question, that pretend^d to unight or rank them with ye planets, having no one particular of common circumstance attending them.

> All the comets have light of their own
> The planets have not
> The comets move nearly in right lines
> The planets move in circles
> The comets are all vast irregu[lar] masses of matter
> The planets are all perfect and regular globes
> The planets are observd to move round [themselves] for diurnal purposes
> The comets are not known to have any such motion or the least use for it.

[27] Lastly many other properties piculiar to these arratic bodies with a due attention to ye position of ye Earth, in its annual orbit, with regard to the point of the heavens they appear in first & last will clearly be account for, besides many other circumstances attending them near the Sun that have frequently occasiond the mistaking of one comet for another, and sometimes 2 distinct comets for one and ye same as pritended in the year 1407 and 1618, all which phenomena with infinite other varigations which have in different ages of ye world been exhibited and strongly 'tis hear presum^d will corroborate the proofs & principals of this new theory.

Besides the above phenomena of ye comet of 1680, if the accounts we have of them may be credeted, other proofs may be also producd, as that of their nature and composition, to induce us to beleive them to be found of a combustable mater, & capable of explosion, even in ther orbits, so as to cause a visible alteration in their most neighbouring trajectories for that of the year 371 is said at length to have broke or parted into two stars, and one in the year 1103 is represented to have chang'd its place by severl leaps & of a very sensible *intersteti*. [28] I shall not desend so low as to account for ye same oporation in booms and water rockets but can't help observing, that subdevided rods of any combustible matter is undenyably capable of such effects & operations.

To conclude: all the new extinct and vanishing stars, together with the cloudy and other variable stars, that are frequently perceived in ye celestial regions, will by this hypothesis be naturally and almost incontestibly accounted for, and without any of these lame and forcd conjectors which astronomers have been reduced to in attemping to explane them.

The first will be found to be no other than fres[h] erruption, or ye revival of an old vulcano, the second such inflamatory substances totally extinguish^d, the cloudy such as are more invelop^d by circumambient vapour, and the variable, with all periodical ones, no other than repeated erruptions of ye same star at different intervals & sometimes nearly at ye same.

The Via Lactea and all other luminous spaces will likewise come under the same definitions as being universally subject to ye same laws & solution.

[29⁴⁹] Plate ye III is designd as an illustration of this theory and as exemplified in the comet of 1680. This projection is design'd to demonstrate that ye phenomena of that comet is equally solv'd by this construction, thô upon different principals than those advancd by Sr Isaac Newton, and founded upon the same observations. The material difference will be found chiefly [in] regard to its laws of motion. That is, it is supposed to have been retarded in its approximation to the Sun and equally to have been accellerated in its return from it, all of which appears to be as evedently demonstrable as that of the increadible velocity a[ssi]gned by the Newtonian principals, which oblig'd it to pass thrô such an intollerable d[e]gree of heat.

In this scheme it may be observ'd that the lines of position or observation are represented as comon to both trajectories and proportionable everywhere regarding equal intervals of temporal motion and the velocites generated in both tracts. That of Sr Isaac^s is sufficiently distinguish^d by its vicinity to ye Sun, and the dotted curve, shews the path it must have pursued by this new theory. By which it appears that instead of keeping so very near the Sun, it could possibly advance no further than betwixt ye orbits of Mercury & Venus. [30] In like maner the trajectories of all the rest of ye comets as represented in Mr Whistons scheme,⁵⁰ the

---

[49] *This page is an insertion, as the original numbering shows.*
[50] *W. Whiston, 'A scheme of the Solar System; with the Orbits of 21 Comets', in a large sheet, engraved on copper by Mr Senex (London, 1712).*

principle of which are shewn in Plate ye I, must consequently be chang'd into a new direction. But as none of them by this theory can possibly return again, it is of no great consequence to construct their respective hyperbolic curves unless it were to find the respective star, or determin the particular point of heaven from whence they were projected.

For[51] instance or a more familiar illustration. In plate ye II:—

Let the lines of position AI; BH; CG; EF present ye observations of the formentiond comet in its approach to ye Sun, and ye respective positions FG, GH, & HI in one with 4,3, 3,2, & 2,1 in ye other, will mutually repr[esent] the intermediate parts in each trajectory as gener[ated] by different principals of motion and so as to determine in ye first case the body to have mov^d in a parabola and in ye 2^d to have describ'd ye curve of an hyperbola.

[31[52]] Now as this is an ambiguous case & like ma[n]y others in mathematics capable of a double answer, and in which we have nothing but ye light of reason to discover ye truth by, I think in our solution of all such unlimeted problems, we aught to adhear to what is most natural & subject to ye least objection, and therefore we may venture to conclude that the last is ye least absurd of the two conic sections, & considering the almost incredible consequences of ye first must have been the true trajectory of this comet. [31'] Thô comets or new stars may be ejected perpendicular[ly] from their respective celestial vulcanos and appear only to change the magnitude for some time as moving in a right line towards the Sun or Earth, in which case they will appear to have no tails as many of them have done, & particularly that of the year $160\frac{4}{5}$ which was observed by Kepler in Serpentarius, yet both in advancing and receding at any considerable distance from y^e stary firmament will still appear to change their geocentric places very sensibly, and particularly the said comet (which was called a new star) did & equall to[53]

This will clearly be demonstrated by ye following diagram.

PLATE III

Let S represent ye Sun, EI the Orbis Magnus and GAH a great circle in ye heavens. Let F be a comet advancing without a visible tail or directly towards the Sun or Earth. The Earth being at I the said comet will appear at A, but when y^e Earth is mov'd to E

---

[51] *Probably another afterthought.*
[52] *This may be a continuation from p. 29.*
[53] *The sentence breaks off.*

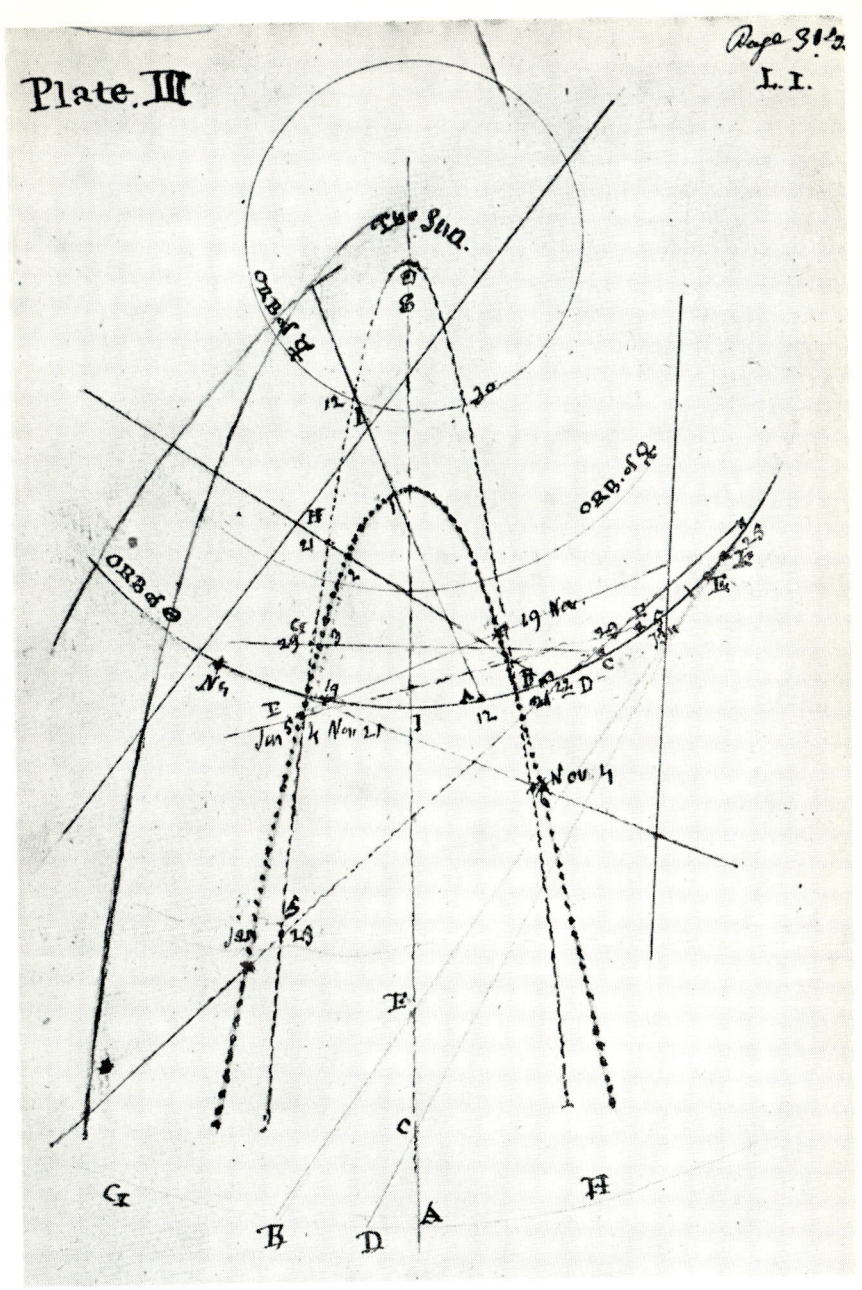

Plate III: the sole known survivor of Wright's sketches for *Second Thoughts*.

and y^e comet to C the said comet will be seen as at B & consequently to have mov'd from A to B: the same is to be understood of its receding, for if first seen at C [31″] and without a train, the Earth being then at I; when ye Earth is advanced to E and ye same comet retir'd to F it will appear as at D, & possibly if not at a very great distance, with some little tail. It is further to be observ'd that ye apparent change of place in ye same comet seen at F or C, when ye Earth is respectivly at I or E, will be greater in advancing, than in retiring from ye Sun. Q.E.D.

But all celestial vulcanos, that only blaze forth a fresh as original or as renewing their extinguish^d fires, without any other eruption of solid mater, ejected to any distance, can only appear as new stars, & when variing their magnituds most can only be consider'd as in a state of less or more intense flame from its quantity of combustible matter.

[32[54]] Thus it follows then S^r as a most rational result, not withstanding the astonishing and almost increadible idea excited in minute minds, that the comets and new stars &c may prove to be no other than enormous siderial erruptions project^d with vast velocity and rapid force thrô the immense aetherial medium from the inflam'd firmament and solid concave regions of the heavens or some other fountain yet undiscover^d and that such as are not absorb'd in ye Suns body as a providential fuel[55] ordain'd to fead his ever lasting or perpetual fire all or most of them must[56] return again with like velocity to their original heaven or subcelestial sphere of the fix^d stars.

It has been clearly evinced to be repugnant to reason and contradictory to the known laws of all projectiles, to suppose that any of them could possibly pass beyond or round ye Sun, consequently they can have no natural connection with ye sensible movements of the solar system. Therefore [they] must only be look'd upon as occasional vissitants within the planetary regions, and hence no doubt ordaind for some wise purpose of the Divine Nature as agents of his infinite power. For besides preserving in full energy the fountain of light and heat, the visible prolific agent & progenitor of all animal & vegitive life [33] the nurse of all nature, and preserver of creation, as alarming and terific spectors

---

[54] *The handwriting and the style suggest that pp. 32 and 33 are of a piece with the opening page of this Letter.*
[55] *Cf.* Newton, Principia, 2nd ed., p. 481.
[56] '*or most of them must*': *an insertion, perhaps related to the suggestion which follows on p. 34.*

they may possibly be generated to ye end that all rational beings may be awakend into reflection and put in mind of the mortal state of their humanity, and our visible dependence upon that sensible power which not only manefestly governs very remote, but also superior states and most undoubtedly and infinitely to all immortal beings in a more glorious & happy state in future life. In this light can we look upon them less than ye merciful minesters & monitors of the Divine Goodness, to beings in a dark & mortal existence, putting us in mind not only of our duty but also of a final desolution of all visible nature, giving us also a capacity to be convinc'd of his illimitable power in order to render us worthy of those eternal immortal joy[s] he is eternally decreeing for all his favourite beings.

Plate[57] y^e IIII represents a section of the creation according to this hypotheses, and if we would form a tolerable notion of the least possible extent of it; let us suppose the greatest annual parallax of y^e fix'd stars to be no less than one minute.

Then will radius — 10.0000000
Less sign 30″ = 6.1626961
Equal y^e radius of y^e magnus orbis } 3.8373039 = in numbers 6875.5

which multiplied by y^e mean radius 80 million and y^e least distance of y^e stars will be found to be
550.040.000.000 miles.

[34] That comets are capable of distroying such worlds as may chance to fall in their way, is, from their vast magnitude, velocity, firey substance, not at all to be doubted, and it is more than probable from the great and unoccupied distance betwixt ye planet Mars and Jupiter some world may have met with such a final dissolution.[58]

---

[57] *An insertion squeezed into the available space. The calculation duplicates that on p. 4 of Letter II below. On Wright's various figures for the distances of the Sun and the stars, see the editor's Introduction. 80 million miles is the figure Wright adopts for the distance of the Sun in* Original Theory, *p. 71.*

[58] *A very remarkable anticipation of developments that were to take place in astronomy after Wright's death. Kepler, in his* Mysterium Cosmographicum, *had considered the possibility of an unknown planet in orbit in the large gap between Mars and Jupiter (Praefatio ad lectorem: ' Inter Jovem et Martem interposui planetam',* Gesammelte Werke *VIII (Munich, 1963), p. 24), and the size of the gap had been remarked upon by other writers. But the idea of a ' missing ' planet came to the fore in the late eighteenth century through the work of J. D. Titius, who*

But as the comets are confin'd to no particular direction, but have the whole sphere of heaven to range in, there is little more than a mear possibility of such dreadful catastrophes ever happening, yet if more of these bodies were seen to move within ye limits of the zodiac[59] and near the plane of the ecliptic they would be very alarming visitants; but as universal bodies and every where directed to ye Sun it is most probable that their sole or chief end is to supply the Sun with fresh and perpetual fuel, as many of them no doubt that reach his inflam'd atmosphere are torn to pieces there, and fall in to his ever burning lava. This manifestly appears from the many fragments of such dissipated bodies, visible to ye naked eye upon ye Sun disk, and flying round him in ye direction of his equatorial parts.[60]

[35 blank]

[36] Here it may be very proper to observe that as no more regular bodies can possibly move round a ce[n]teral one thô in all maner of directions than can be describ'd upon any one plain of ye same radius, it naturally follows that all the planets may have originally & precisely mov'd in one and ye same plaine, since it is most evident from such a primary disposition, [that] the danger of being disturb^d by other eccentric bodies, would be reduced to its very least, or naissant degree, and in this the Wisdom of Providence appears to be no less, than in the formation & disposition of ye whole system. For it is highly probable that all the different inclinations and various obliquities of the planets motion, have from time to time arisen, in consequence of ye near approach of some of the greater

---

pointed out in 1772 that the known planetary distances from the Sun are nearly in the ratio of the numbers 4, 4+3, 1+2.3, 4+4.3, 4+16.3, 4+32.3, and that the sequence lacks the term 4+8.3. This curious relationship, which Kepler himself had only just missed, greatly impressed the young J. E. Bode, and the discovery of Uranus by W. Herschel in 1781 at a distance close to the next term in the sequence, 4+64.3, forced astronomers to take the 'law' seriously. An organised search for the 'missing' planet was eventually instituted at Lilienthal in 1800, but it was G. Piazzi at Palermo who in 1801 chanced upon what proved to be a very small planet (or 'asteroid'), Ceres, in orbit between Mars and Jupiter. The following year a second, Pallas, was discovered by W. Olbers, who suggested that these were two fragments of a planet which had been blown to pieces either by internal forces or by the impact of a comet, and rightly forecast that many more such fragments would be found (see Morton Grosser, The Discovery of Neptune (Cambridge, Mass., 1962), pp. 27–38).

[59] Cf. Newton, Principia, 2nd ed., p. 480, and J. D. Cassini, 'Première idée d'un zodiaque des cometes', Histoire de l'Academie royale des sciences 1 (1733), p. 160.
[60] Cf. Clavis Cœlestis, p. 49. For contemporary explanations of sun-spots, see Roger Long, Astronomy 2, Book 3 (Cambridge, 1764), pp. 478–9.

come[t]s, in their vagrant way thrô the mundain space.[61] And hence also must inevitably arise considerable errours, and incorrectness in the theories of all such planets, that have been observ'd in ye too near neighbourhood of any such approximating comet, and which will consequently require in time to be new rectified.

---

[61] *Whiston* (A New Theory of the Earth, *pp. 117–18*) *ascribes to the passage of comets the deviations of the planetary orbits from coplanar circles.*

LETTER THE II OR 11[62]

Sir

As the end of all human observation, study and experience is manifestly to promote in us a just knowledge of our own existence and situation in life and a right notion of the natural dependence w[e] have upon the indulgent and divine Author of our existence, I presume it will not be unacceptable to you, to give you my thoughts & ideas upon this subject so far as regards the final causes and sensible structor of the visible universe or creation in general. [3] And[63] first in order to have as clear and full an idea of the universe as possible, let us imagine Plate ye IIII to represent a semisection of all the known creation as far as the solid firmament of heaven, or visible concave region of the stars.

In this scheme the spherical body in the center represents the Sun, and the concentric circles round it such orbits of the regular planets as have already, or are yet to be discover'd; for I am far from supposing our present knowledge of ye solar system perfect and fully known.[64]

The parabolic & hyperbolic curves will then also justly represent the various trajectories of all such comets as have any time made their respective appearance, and become visible to us on ye Earth, and is evedently ejected from their fiery fountains, or celestial vulcanos which consolidate & form ye stary firmament.*

[1 cont.] The free and uninterrupted passage of the comets to and

---

[4] * If we would form any tolerable notion of the extent of this vast sphere of visible existence, and according to this hypotheses, let us suppose the greates[t] annual parallax of the fix'd stars which are here to be conceiv'd all equal, to be no less than one minute of a degree.

Then will radius — 10.0000000
Less signe 30″: — 6.1626961

Equall the radius of the Magnus Orbis } 3.837,3039 = in numbers 6875.5

which multiply^d by y^e mean radius 80 million miles will give y^e least distance of the stars, to be 550.040.000.000 [miles].

---

[62] *The original title, barely legible, appears to have been:* 'A postscript ... to the last letter of the Theory of the Universe'. *The manuscript, though considerably revised, is much more of a unity than that of the previous Letter, and pp. 1–26 include sheets originally labelled* 'A', 'B', ..., 'S'.

[63] *This sentence and the two following paragraphs are a later insertion.*

[64] *Earlier references, implicit or explicit, to the possibility of there being undiscovered planets include:* W. Whiston, Astronomical Principles of Religion (*London, 1717*), *pp. 15, 19;* W. Wall's *contribution to the second edition of Tobias Swinden's* An Enquiry into the Nature and Place of Hell (*London, 1727*), *p. 355;* T. Wright, Clavis Cœlestis, *pp. 16, 17, 33, and* Original Theory, *p. 31; and Kant in* W. Hastie, Kant's Cosmogony (*Glasgow, 1900*), *pp. 65 seqq. One planet was discovered in Wright's lifetime: Uranus, by William Herschel in 1781.*

from the Sun as in perpetual intercourse betwixt the visible regions of the stars & planetary orbits evedently demonstrate a natural connection betwixt ye whole and all its parts so as to reduce to reason the absolute and necessary idea of ye great unity of those infinite attributes that constitute the Deity, and his eternal dominion and all alike consider'd whither with respect to ye existence of matter, duration and space, as all animal & vegetative life exists by a regular circulation of material or mineral fluids, so, may universal nature [2] be also sustained & preserv'd by a like circulation of a more refin'd & subtile celestial mater, and next to spirit.

Some of the natural consequences attending such a construction of ye celestial regions as in ye forgoing pages I have there indevoured to demonstrate may be presum'd to be these:
First

The inhabitants if any such there be, may be indulg'd with more[65] fix'd seasons and a perpetual or eternal day; they may also have a proper deversity of heat and cold and of gloom and shade, though they can possibly have no real night; but what advantages would or could arise in such a dispensation of nature, it would be judg'd impertinent for human philosophy to indeavour to explain.
Secondly

They may have likewise and very possibly much larger tracts of teraqueous regions not only seperated from each other by the impassible and impenetrable rounds, but also vast worlds of earth floating upon immense oceans, to which our Earth would only appear as a small rock or little island, what rivers? what woods? what plains & mountains may not be expect'd to fill & adorn so illimitable a sphere. [4] Supposing this vast concave chaos, and the starry regions heaven; though Milton's very imperfect notion of it was far from ye truth yet how naturally will it illustrate his fine ideas on the subject; and as juditially put into the mouth of his first apostate

> O! thou that with surpassing glory crown'd,
> Look'st from thy sole dominion like the god
> Of this new world; at whose sight all the stars,
> Hide their deminish^d heads: to thee I call,
> But with no friendly voice, and add thy name
> O Sun! to tell thee how I hate thy beams,
> That bring to my remembrance what I was
> From what height fallen.[66]

[65] '*more*': *an insertion in the original draft.*
[66] Paradise Lost, *Book IV, ll. 32 seq.   The usual ending is:* 'That bring to my remembrance from what state | I fell.'

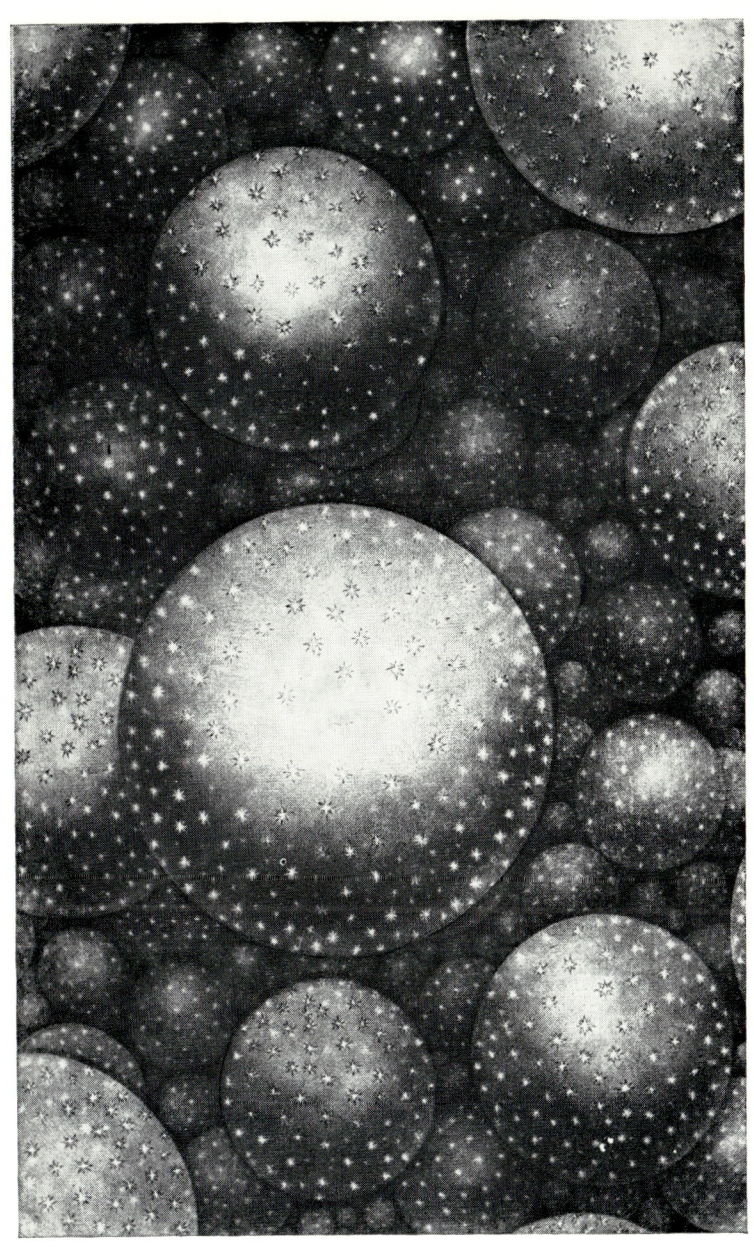

Plate XXXI of *An Original Theory* (see page 49).

of our own solar system: so that every star may not only be the radient center of a seperate world of creation but also a mutual luminary, in common with the rest, to all the neighbouring & bordering regions, and hence the desolation, arising from eruptions, can be no more to them, then those of burning mountains upon Earth to us.

The progressive measure of time to them must also be distinguist or mark^d by much larger intervals or periods of duration, [8] proportion^d perhaps to our sensations of it as human beings to ye life of an insect, counting their days or hours only by erruptions & returns of comets and new stars, which probably cooling again before they reach ye celestial reagions may there break into an infinity of jewe[l]s as the ultimate and most intense effect of fire. Thus every erruption by the wise Creator may be ordained without digging into the bowels of the celestial firmament as we do upon Earth, without labour to be perpetual fountains of celestial treasure not only of every kind of precious mettle but also every sort of precious stone infinitely more deversified than all or any of our Earths production are, totally unknow[n] perhaps to human natures and only as ye produce & eternal wealth of heaven. Knowledge there may also pertake of like degrees of perfection and all their ideas and tast may be inlarg'd in adequate proportion. What mansions there may not be erected, and what infinite ornaments of distinction may not adorn such beings & dignifie their celestial and more devine natures. [9] The maner of existing here perhaps, may be very different to that we now enjoy upon Earth and are sensible of in our present state.

The vast extent of these indefinite regions, which comparatively with any portion of finite space we can possibly have any adequate notion of, may phisically be considerd, if not term'd, infinite; and consequently a being existing in these unlimited realms, may probably be endowed with locomotive power equal to ye full enjoyment of this celestial universe.

That is to say, as the human imagination here can pas[s] from object to object as in an instant and rise up in idea what ever is most pleasing to ye mind, the immortal spirit in the ethereal or immaterial state, may have a power of putting on not only what form or shape it pleases, but also have the facculty of moving from place to place with like ordaind facility, and instantaniously. In short to have whatever it pleases, when it pleases & where it pleases all in the utmost perfection & subject to its own absolute & incontrollable will.* And if so we may as naturally conclude that the

* Se[e] Spectator N°. 600. [By Addison]

LETTER II 47

[5] When we consider those vast immeasurable regions, all crowded with vulcanos and surrounded with floods of fire, many of whose central craters are much larger than the Sun, and emitting rivers of burning lava too large for our imagination to reach, what less then infinite Power could produce them? Or less then infinite Wisdom comprehend them? Or what less then Immortal Natures can inhabit these? For us nothing is ordain'd or made in vain, those radiant realms of everlasting glory, must not be imagined as a vacant desolation, void of all beings.

In another place[67] I have indeavoured to discribe what various worlds, might be imagin'd to people the Univers, as so many planetary bodies, revolving round the stars, and as all consider'd as the several centers of radiant systems. But here a much wider field is disclos'd to ye imagination and all in one infinitely expanded surface.

What therefore has been said upon those planetary worlds will naturally hear be far surpast. [6] And principally as being ye habitation of an undoubted rase of superior beings. But hear I can proceed no further, having no adequate conception to lead me on, and must of force conclude, borrowing the words of St Paul upon a like occasion

> The eye hath not seen, nor ear heard,
> neither hath it enterd into the heart of man,
> the things that God hath prepared for them that love him.
> [1st] Corinthians II. v.9.

So also Isaiah

> For since the begining of the world, men have not heard, nor perceived by ye ear, neither hath ye eye seen or God besides thee, what he hath prepared for him that waiteth for him.
> Isaiah: 64.4.

[7] Thirdly, their air may be of so dense a nature as I have in an other place observ'd, that flying may be as natural & easy an exploration their, as swimming hear with us on our globe.

If any objection arises to the apparent vicenity of the eruptive regions or starry volcanos as imagining the whole cope of heaven to be one vast valt of fire, and that no such vacant habitable realms may probably be found: give me leave to put you in mind that the most neighbouring of ye smallest stars even in the Via Lactea, may be demonstrated to be at least as far asunder, as the greatest radius

---

[67] *E.g.* Original Theory, *p. 76.*

more knowledge we acquire in our present state the more able we shall be to *will* or evite[68] our future objects of happiness. [10] If by any tracts of reasoning we can bring our selves to believe that this is nearly the construction and state of the glorious subcelestial concave, what may we not imagin the external superficies or convex sphere of that prodigious orb to be, expos'd to the unbounded infinite. If we suffer our minds to rest upon the finity of one creation, here must of necessity commence an eternal and immeasured darkness void of all matter motion & existence. But this our admitted and universally allow'd ideas of the devine attributes, will be no means allow us to suppose. It must then naturally & most necessaryly resu[l]t, either into another superior solar body as at *A* Plate ye IV[69] to enlighten a more magnificent new and semilar creation, or prove an infinitely more perfect world, or new Earth subject to ye laws of a more durable, and devine system of superior planets as suppos'd at *B* of the same figure etc.[70] Thus orbs inorb'd, or by a concatination of orbs successivly inspherd within each other the solid substitutions or modified ideas of ye Devine Nature may be conciv'd to fill all space with its activity and infinitely [11] deversified with restless oporations, perpetually to exert in both all endlessly, and co-eternally with him self.

But this must I fear alway remain an insolvable doubt to mortal natures, namely whither the universal creation is thus composd of one only infinite exten[s]ion of concentric states & beings shell within shell & orb within orb whose center, and also its circumference is equally inapproachable as represented in Plate V, or an infinity of finite creation[s] constituting a like endless plenum as represented in Plate ye 31 of the *Theory*, every where there being equally the center and circumference of all nature: one of these the author humbly conjectors may more than probably be the real case, but here he owns and confesses his most expanded ideas are all absorb'd and lost in the immeasurable powers and capacity of the Divine Nature and concluding it no adequate task for any human being, must here resign the subject as he found it all incomprehensible.

[12] To what has been said may not improperly be added other probable circumstances that may prove natural enough to such celestial regions viz. that those siderial mansions or habitable realms of beings, as existing in an indefinite concave cannot be in all things

---

[68] *Evite = avoid.*
[69] *This and the following references to Plates IV and V have been added later.*
[70] *Originally, ' devine system of beings '.*

similar to ours, for instead [of] a finite horizon with us & of no great distance from y^e eye arising from the convexity of the Earth, they every where must be surrounded with a kind of aura etheria in all ye opposite parts to their central luminary, which together with the radiations of an infinity of circumbounding stars must in effect produce to them an en[d]less day.

This sort of perpetual aurora must far surpass in luce[ncy] & glory any thing we can have any notion of upon Earth, unless we submit to borrow ideas from the *prism*, which is experimentally productive of colours far surpassing even that of ye rainbow. Add to these the periodical and frequent irruption of ye comets & new stars, which are constantly playing against their zenith as doing a sort of homage to ye central Sun [13] and deversifiing that awful and replended body near which no stars can possibly appear and preserve their lustre as here in ye absence of that all glorius luminary, the Moon with us supplys and chears our nights.

In a word the infinite powers of Nature under ye direction of a Providence most supremely wise are so unlimitedly great, nothing that human wisdom can conceive must or may be expected to arise to any comparative degree of the Devine Perfections. But as the natural & experimental philosophy in subjection to our own reason teaches us many things which without such aid could never rise in our ideas, we have a natural right inherent to our senses, to examine every circumstance connected with our being and as wh[e]ther possible or probably in any future state to be expected. For [14] as all parts of the Earth its self which is but as a point to ye rest of ye solar system is not yet known to all its inhabitants, and by much the greatest part of it not yet discover'd, why are we to suppose that our knowledge of an indefinite system of cœlestial bodies can be sooner perfected and completed which extends beyond ye sensibility of all material forms as imagnifiable objects.

Is it not more reasonable to imagine a cœlum in cognito beyond ye known planets than to suppose a terra in cognito at present upon Earth. It is therefor my opinion, that there are or may be, many more bodies belonging to the system of ye Sun, whose more feeble light has not been able yet to reach us at ye Earth, besides others perhaps within ye orbit of Mercury, though lost or lying hid to us in a too radiant state of light.[71]

[71] *Since Wright's death have been discovered Uranus, Neptune and Pluto, all beyond Saturn, and the asteroids or minor planets, which occur especially between Mars and Jupiter. It is believed that no intra-Mercurial planet exists; 'Vulcan' was postulated to account for the advance of the perihelion of Mercury, and supposedly observed, but the advance was subsequently explained by Einstein without recourse to an unknown planet.*

[14ᵛ] When we consider the almost infinite disproportion[72] betwixt the starry reagions and that of yᵉ Sun, the one being to yᵉ other not less than 1,000,000,000,000 to 1, we can not help admiring with yᵉ utmost astonishment, the infinite wisdom & power of yᵉ Creator in so suiting beings of a rational nature to yᵉ latter, whose sensation of magnitude, virtually considered, renders their diminutive state in the vast ideas of space, equal to yᵉ former: in so much that some philosophers have been willing to reject both matter & extension out of Creation and there is no way absolutely necessary towards producing the universe out of nothing, but as only & eternally existing in the ideas of the universal mind.

[15] The ancients especially the philosophers seldom dismis'd a subject of any consequence without considering it in a universal view, and to what end it most obviously was ordain'd. In this light the caballistic writers of the Middle Ages have supply'd the regions above the fixed stars and without the Primum Mobili with a subordination of states, as all inspher'd with in each other and under their respective hierarchys & as all subject to the one infinite eternal Being of Beings.

Docter Flood in his metaphisicial works[73] upon this topic has rang'd them in the following order, and first as the most elevated or superiour he supposes the Seraphim, providential agents or gardian spirits over yᵉ Cherubim;

secondly the Cherubim, having a like charge over the Throni;
thirdly the Throni, as inspectors of the Dominationes;
fourthly yᵉ Dominationes, as intelligences to the Potentates;
fifthly the Potentates, as geni to yᵉ Principates;
sixthly yᵉ Principates, as tutilaries to yᵉ Virtu[t]es;
seventhly yᵉ Virtutes, as influencing the Archangeli;
eighthly yᵉ Archangeli, as superintenance to yᵉ Angels; and
lastly yᵉ Angeli, as guardian agents of human natures in yᵉ regions of mortality.

[16] But as these are not very easily comprehended and more imaginery than manifest beings, I shall make no further remarks

---

[72] *In* Clavis Cœlestis, *p. 5,* Wright *adopts* Whiston's *figure of 763,000 miles for the solar diameter. If the radius of the Celestial sphere is, as estimated in this Letter, at least 550,040,000,000 miles, then the surface of the celestial sphere is more than 10^{12} times larger than the surface of the Sun.*

[73] *Robert Fludd,* Tomus Secundus de Microcosmi Historia *(Oppenheim, 1619), frontispiece and Tr. 1, Sect. 1, Lib. 1, p. 37. Fludd's work was little known in England in Wright's day and so, until the discovery of this proof that Wright was (at least in later life) acquainted with Fludd, the suggestion by Professor Walter Pagel that Fludd might be the source of Wright's 1734 ' outer darkness' seemed implausible.*

upon them hear but refer the reader to that ellaborate worke printed in ye year 1619.

From this order of being no doubt Milton deriv'd his great ideas, in the address's his Satan makes to ye fallen angels:

Thrones, Dominations, Prinsdoms, Virtues, Powers.[74]
Again:
Powers and Dominions, Deities of Heven.[75]
And further:
Thrones and Imperial Powers, offspring of Heaven,
Ethereal Vertues.[76]

But the same might be gatherd also from St Paul's Epistle to the Colossians who seems to be well acquainted with all ye ancient learning of the Jews and Caldeans etc.:

For by Him (i.e. God) were all things created that are in Heaven, and that are in Earth visible or invisible whither they be Thrones, or Dominions or Principalities or Powers.[77]

[17] The Romans rose in their ideas of a future state of existence no higher than the visible starry regions, which they supposed to be the habitable mantions of all their Heros, Philosophers and Demigods. This is no way repugnant to the present theory.

There those whom wisdom hath exalted, shine
Just Solon, stout Lycurgus, the Divine,
Plato, and he who made him such whose doom,
Justlier condemns his Athens; he by whom
Persia was foyl^d, who strow'd with fleets the main
And Roman worthies [the] more numerous train.

And farther:

More High th' immortal gods have fix'd their seat
Next whome is this, [with] god like men repleat.[78]

Manilious in his learned work has taken occasion to innumerate all the most illustrious names antiquity could furnish him with, as the comōn inhabitants of these immortal reagions; which for a more full account of I must direct the reader to his Poem on the *Sphere*.

[18] In the Christion despensation, Josepheus is the only author I can recollect that has ever attempted a description of a local or religious Hades, which he has given us in an Homili suppos'd to

---

[74] *Paradise Lost, Book V, l.772 or 840.*
[75] Ibid., *Book II, l.11.*
[76] Ibid., *Book II, ll.310–11.*
[77] *Col. 1 : 16.*
[78] *Edward Sherburne*, The Sphere of Marcus Manilius (*London, 1675*), *pp. 52, 58.*

have been wrote when he was Bishop of Jerusalem,* and probably erected upon the out lines of such a state as given in ye Revelation of St John and hinted at also by St Paul in his Epistle to the Thessalonians.

Sir Isaac Newton, not knowing where to fix or find a local residence for the sole Supreem Being or Infinite Nature has contented him self[80] with ennumerating some of the principle Divine attributes** and leving us in all things else to be lost in a kind of infinite common wealth of nature.

[Pages **19, 20** and **21** missing]

[**22**] Thus by means of a multiplicity of comets in all mañer of directions a Providential Fire is made to circulate thrô all the frame of Universal Nature and like the blood of animals which nurrishes and feeds all vital beings, purifies the etherial medium and probably supplies all its central suns[81] with fuel.

Every ordination of beings being a perfect creation of itself, and independent of all others having its own respective past, present, and future state piculiar to itself, but yet is so dependant upon the whole that if any one of ye concentring fabrics faild, all the rest would be involv'd in total desolation. But the infinite continuation of ye universe is such and the support of it so natural that this in evry part of its unity appears to be absolutely impossible & consequently it must be eternal. [**23**] The solar body it self in evry system purifies its own planetary region and the comets and eruptive stars perform that salutary office in ye more remote siderial regions of evry creation and hence it is perhaps, that more frequent ejections are requir'd in that cœlestial aether, not only upon account of its indefinite expanse, but as being in a maner

---

* Se[e] Mr Whiston's Evidences and translations of that author.[79]
** In which he seemd to have follow'd Milton's ideas of ye nature of angels. See Book VI. 1.350.
    All heart they live, all head, all eye, all ear,
    All intellect, all sense, and as they please,
    They limb themselves, and colour shape and size
    Assume as likes them best, condense or rare.

---

[79] '*An extract out of Josephus's Discourse to the Greeks concerning Hades*', in Works of Flavius Josephus, *trans. by W. Whiston (4 vols., London, 1806), iv, pp. 360–362 [recte: 364]*.
[80] *In the* General Scholium *to the* Principia.
[81] *In this passage Wright discusses the '* concentering fabrics*', so the '* suns *' form an hierarchy, one inside another. To judge from what is said on p. 26 of this Letter, the missing pp. 19–21 may have treated of independent '* creations *' as illustrated in Plate XXXI of the* Original Theory.

excluded from all ye benefits of the central fire; which though to us is the fountain of light, to the inhabitants of ye starry mansions, if any such there be, may be totally invisible*. [24[82]] From the infinite, and eternal attributes of God, in consequence of his absolute and infalible Goodness, Wisdom and Pow'r, by some philosophers it has been infered that his Providence must be an involuntary and not a free agent, the Devine Being itself or naturally existing as from necessity or fate. And in this predicament he cant be suppos'd to act otherwise than he dose as being ye consequence of circumstantial and perfect rectitude of principle or action. Hence it may naturally follow that as an infinite and eternal being, existing by a fatal necessity, all his divine attributes are also as fatal necessities too, and consequently infallible and cannot err, i.e.: as necessarily wise, & necessarily powerful, also necessarily good.

In all probability this consideration may have lead the Gentiles to conceive their Jupiter as subject to their Fates whose decrees or destinies he always ratified but could not alter or control. [25] If creation is therefore perfect it must of necessity always have been so, and will likewise be eternally perfect thrô all time.

Perfection therefore must consist not in the finite mode of its parts which are all perishable but in the excellency of its constitution as arising from the infinite variety and eternal mutibility of the whole.**

[26] To conclude I must now observe that as the construction of the frame of nature must of consequence either be in the form of an infinite concentric unity of states, or of an infinitely expanded disposition of its parts, all independent of each other, one of these modes most probably may be found to be the case and prove the real & existing mode of the creation. But wether ye plurality of systems & multiplicity of worlds as have been represented in the *Theory*, or their united ordination, in infinite gradations, as here endeavourd to be demonstrated, is judg'd to be most probably, the perfect system of creation, or nearest to the universal mode, the

---

* Note: that celestial eruptions may continue a very long time, without any sensible variation, is evedent from terrestrial ones, particuly Stromboli, one of the Lipari Islands, which perpetually emits fire with seldom any lava. *Quer*: An idea of a celestial volcano drawn from ye surpassing [?] flames of the Sun.

** For other better who would worce produce
                                                —Milton

---

[82] *This page and the following have been interpolated.*

LETTER II 55

rationale & unprejudiced reader must take upon himself to determine and reconcile to his own ideas,[83] [27] the author thinking he has done his duty in endeavouring thus far by ye help of science & philosophy to unveil infinity to him and lay the universal scene before him, and much more open here than yet our earthy telescopes have ever reachd. [28] Here it must be observ'd, and as of no small consequence in astronomy, that the stars, being all look'd upon as so many celestial eruptions, promiscuously bursting from the firmament of heaven, and no longer consider'd as radiant suns, illuminating the mundane space, are suppos'd to be originated from a combustible and infinitely various inflamitory substance not unlike the burning mountains upon Earth but of a more subtle nature and like them also have their stated periods of a bright or faint lustre (hitherto inexplicable) varying & changing the position of their radient points: sometimes becoming extinct & again reviving, or new breaking out.

It follows that the orifice or aperture of this starry flame though of no very sensible diameter to us, must be perpetually varying its punctum or place, upon its respective celestial vulcano, like that of *Visuvious* or any other burning mountain; consequently this pyracula or eruption will have a vertigious motion, such as may be supposed in the Newtonian philosophy to attend the Sun, in consequence of his moving round the common center of gravity of ye system,[84] and hence also to be mutually observ'd in all stars, consider'd as solar bodies from one another.

[29] That such a vertigia is not a meer conjector and founded on suspicion, a local motion in a sufficient degree has been observ'd in upwards of 300 stars.* But as it is understood at present to be rather an errour either arising from ye instrument, or the incorrectness of the observer, we must refer the confirmation of it, to further observations of ye most able astronomers and more correct instruments.

* See y^e *Nautical Almanack* for ye year 1773.[85]

---

[83] *The original draft ended here with ' ... his own ideas of the state of Nature'. Wright appears to have realised that the proper or individual motions observed in certain stars provided a powerful argument against his ' volcano' theory.*
 [84] *Newton,* Principia, *Book III, Prop. XII, Th. XII.*
 [85] *'A catalogue of the places of 387 fixed stars ... with their magnitudes and annual variations in right ascension and declination', annexed (pp. 25–38) to* The Nautical Almanac *for the year 1773 (London, 1771). Wright is confusing and probably confused: the ' local motion' has not been ' observ'd', but is simply the standard correction for precession; the danger to Wright's theory lies not here but in the much smaller number of observed motions proper to individual stars.*

But I am well aware that this or a like vertigia may also be attributed to another cause (viz.) from a pendulus libration of ye Earth itself in consequence of the flux and reflux of ye sea as vibrating with ye lunar influence.

But what ever is the true case it manifestly puts a stop to all expectation of ever ascertaining the lunar theory by means of ye stars, to any greater degree of perfection than the present elements of ye Moon has ariv'd at. And hence we may also justly reject all, or most, of the trifling minutiae in modern equations (I mean such as in their maximum only arise to a very few seconds) as entirely useless and but an embarrassment. [30] For if a very few seconds must be consider'd, either in the calculus or in the observatory, many more equations may be produced from the Newtonian principals of gravity, than have ever yet been admitted,[86] particularly that of ye Earth's continual approximation to ye Sun, and the Moons perpetual approximation to ye Earth: with ye Moon and Mars as Panselenes in opposition to the Sun. But I think all these are meerly speculative and can never be reduced to practice.
[31] Lastly the dimensions of ye system or universe in magnitude or extent, are found from ye diurnal parallax of ye Sun, and ye añual parallax of the stars. The first having been determined by ye transits of Venus over ye Sun,[87] as observ'd in differ[en]t parts

---

[86] *Cf. David Gregory*, The Elements of Astronomy *done into English (2 vols., London, 1726), ii, pp. 529–30.*

[87] *On the transits of 1761 and 1769 and their results, see Harry Woolf,* The Transits of Venus *(Princeton, N.J., 1959). Figures derived from the 1761 transit ranged from 8.28″ to 10.60″, while those from the 1769 transit ranged from 8.43″ to 8.80″ (Woolf, op. cit., p. 192). Wright's earlier figure (this Letter, p. 4) of 80 million miles for the distance of the Earth from the Sun corresponds to a parallax of over 10″, and is a value he uses in 1750 (Original Theory, p. 71). Until 1769 it was a defensible estimate (cf. Thomas Hornsby, 'A discourse on the parallax of the Sun',* Philosophical Transactions *53 (1763), pp. 467–95).*

For Whiston's value of 47″, see his Astronomical Principles of Religion *(2nd ed., London, 1725), p. 25. In ascribing the value of 20″ 2‴ to Bradley Wright is committing an astonishing blunder; this figure should refer to the aberration of light in the case of Gamma Draconis, the true parallax being undetectable but less than 2″, as Wright well knew (Epigenoma below, p. 1; James Bradley, 'An Account of a new discovered motion of the fixed stars',* Philosophical Transactions *35 (1727–28), pp. 637–661; cf. Wright, Original Theory, p. 71).*

In the following calculations, Wright computes in footnotes the values of cosec 9″ and cosec 20″ 40‴ to obtain the distances of the Sun (the radius of the orbis magnus) and of the stars (the radius of visible creation). He wishes to compare this last figure with the radius of the Sun, which he obtains from the formula

$$\frac{\cos 16' \ 3''}{R} = \frac{\sin 16' \ 3''}{r},$$

where R is the known radius of the orbis magnus and r is the required radius of the Sun.

LETTER II

of y^e world, and the last by *Gamma*, a star in the Dragon's Head observed to pass the meridian, near the zenith of London.

By these observations the Suns horizontal parallax has been lately deduced to be nearly 9″: and that of a star in y^e ecliptic pole, to y^e radius of y^e orbis magnus: 20″: 40‴ or 43.*

From whence the Earths mean distance from the Sun is determin'd to be nearly 22918_L4** semidiameters of y^e Earth, and the distance of the stars*** 9959_L85 radiuss of the orbis magnus. Viz: The radius of y^e orbis magnus 90,934,878_L375 miles and the radius of y^e universe or visible creation will be 905, 697, 748, 383_L 24375 miles.

[logarithm]

[32] Again, as y^e sine compliment of the ☉ˢ apparent diameter, viz: 9,999996

is to y^e ☉ distance in semidiameters of y^e ⊕ 22918_L5, viz: 4,360183

So is y^e sine of y^e ☉ˢ app[arent] semidiameter 16′ 3″ 7,669200

_____

12,029383

to y^e log of y^e Suns true ½-diameter 2,029387

Viz: 107, in semidiameters of y^e ⊕ and hence $3967.75 \times 107 \times 2 = 849{,}098_L5 =$ the true diameter of y^e Sun in English miles.

N.B.: the increase of all the parts of the rest of y^e system, may be found by the common rules of proportion, and the result will be that the radius of the orbis magnus, in radiuss of the Sun will be 214 and the distance of y^e stars or radius of y^e univers or visible creation will be nearly in round numbers 10,000 radiuss of ye orbis magnus. Hence ye diameter or magnitude of ye visible univers, to the diameter of ye Sun is as 2,140,000 to unity, or 1. [33] And as all spheres are to one another as ye cubes of their diameters, it follows that the magnitude or expance of y^e visible creation or orb

---

* Mr Whiston from y^e observations [of] Hook & Flamsteed made the greatest parallax of y^e stars 47″: but by Molyneux & Bradley's more correct observation it was found to be 20″: 2, nearly as above.

** Sine radius   10
   less sine 9″   5.6398174

= log: distance   4.3601826
  22918_L4   //

*** Sine radius   10
    less sine 20″: 43‴   6.0017470

= log: distance   3.9982530
  9959,85   //

of y^e univers, as bounded by y^e starry firmament, will be to y^e globe of y^e Sun as $2140000 \times 2140000 \times 2140000$, or as
$$9,800,344,000,000,000,000 \text{ to } 1,$$
a most astonishing and almost indefinite proportion. But to reduce this prodigious comparison in some measure within the power of our reason & as an object of sensation, let us imagin one pepper corn of which 10 will make one inch to represent y^e Sun, and then it will be found that an orb or sphere comprehending all ye sensible universe in like proportion, will take up 17833 yards or upwards of 10 miles diameter. Thus visible matter compound with its sensorium, space, is most rationally reduced to little more than nothing and as its smallest particle as an object is the indivisible, what can its ens primum proceed from, but an inarchable and unutterable and incomprehensible Being.

[34] The stoics, Zeno & Heraclittus afirm'd God to be an immortal, rational, and perfect Being, consious of his own happiness, subject to no evil, and govern'd the universe, and all things in it, by his absolute and infallible Providence.

But at ye same time they are said to have held two principals *viz*: God and Matter, the one active and the other passive: and both corporeal, for they had no idea of an incorporial substance: and hen[c]e in effect their *dual* principal must be resolv'd into a unity or one. This infinite Power, was the Father of ye universe, and their Divine Architect of creation.

They also conceiv'd this Divine Nature to be a firy spirit, void of all figure, and as existing in a state of mutibility, with evry power of modification, & [it] methodically generated or produced the world in all perfection, according to ye laws of fate or nature.

[35] This sect has been accusd of confounding God and Nature to gether, but it rather appears as a more rational conclusion that there God was the soul of [the] world, and hence the Pythagorian philosophy as given us by Virgel. Even in the Dark Ages of atheism & infidelity were all inducements to them to think that,

>All things are but alter'd, nothing dyes
>And here and there the unbodied spirit flys
>Till roving round, again new limbs it finds
>And actually those according to their kinds.
>*Then let not piety be put to flight.* &c.

The Desidets[88] or Indian Jouties and Persian Suffis were also of this opinion, beleving the Divine Nature as immovable, & un-

---

[88] 'Desidets': perhaps Wright intends 'Docetists'. 'Jouties': cf. 'jaties' (*Bengali*) = *astrologer, astronomer.*

changeable in attributes, produced both spirit & matter, out of his own substance, as the efficient cause both of extraction & extension & also of resumption in him self again, as y^e act of desolution. But as their alligorys are not equal to y^e subject I shall say no more of them in this place.

[35.2] Josephus in his lear[ne]d book against Apion tells us their Divine Legislator taught them to regard God as the author of all good things that were enjoy'd either in common by all mankind or by each one in particular, and that it was impossible to escape God's observation even in any of our outward actions or in any of our inward thoughts.

He also represented God as unbegotten and immutible, thrô all eternity: superior to all mortal conceptions in *pulchritude: and thô known to us by his power, yet unknown to us as to his essence.

The first command is concerning God: and affirms that God contains all things; and is a Being every way perfect, and happy—self-sufficient, and supplying all other beings; the beginning; the middle & ye end of all things, He is manefest in his works & benefits; and more conspicuous, than any other being what ever, but as to his form and magnitude. [36] Thus most naturally the Sun as centered in a sphere of glory; thô not as the Divine Nature its self; may justly be lookd upon as his visible throne [even] thô otherwise vacant of his personal presence, since all creation is full of it. Here his allseeing Eye of Providence is more imediatly fix't, where viewing all ye visible creation, and with unlimited sensation & perfect knowledge, He dispenses i[n]falibly good & evel, according to his unalterable laws & eternal justice.

But the Sun is neither God nor an object of worship, thô it is natural & can be no impiety to turn our faces towards it as the most glorious object of ye visible creation, when we meditate on the divine excellence, or contemplate any of its infinite & eternal attributes. The infinite unity, or all nature united in its immutible Godhead, may be a very rational object of adoration, but their can be no one finite object of worship without *idolatry*.

---

* This may have not been unknown to Plato & probably lead him to his Supreem Beauty.

LETTER YE III OR 12[89]
upon
the infinite unity

Of systems possible if 'tis confest
That wisdom infinite must form the *best*
Where all must *fall* or not coherent be:
And all that rises, rise in due *degree*;
Then in the scale of life and sense, 'tis plain
There must be *somewhere*, such a rank as man.

[1] As in our present state of being we have a capacity of contemplating a probable inferiour state of existence as in the body of the Sun, and also a like power of considering the existence of a much superiour order of beings in the super coelestial regions, far without the visible starry firmament, so may we naturally suppose that those beings within ye radiant microcosm of ye Sun, likewise have their adequate and proportional powers, adapted to the proper purpose & contemplation of other beings, as much inferiour to *them*, as they are most manefestly inferiour to *us*; and that beings also infinitely above us may still have superior faculties of sense & reason to regard and contemplate other creations in a like ratio, as far above them, and at the same time may be endow'd with a capacity of reflecting upon us in our more minute state, the Divine Nature and his radiant presence being every where magnified with his visible works, and as appearing always & every where in adequate glory, with his present exhibition of things or state of nature in every visible sensorium[90] or sphere of his creation. [2[91]]

PLATE V

[89] *The title page has been much altered. The original title included the words* ' General Scholia Concerning Unity ', *and the original mottoes were, first,* ' The Sun is Born ', *and then* 'All is full of Jove '. *The Letter, like its predecessors, is a patchwork of separate pieces. Pp. 1-20 appear to comprise fifteen pages of an early draft, with additions. The* ' Scholia ', *pp. 21-23, and its continuation, pp. 25-28, are two further components, while pp. 29-31 of* 'Final Idea or Scholia ye 2d', *were originally numbered pp. 9-11; pp. 32, 33 are recto and verso of yet another sheet.*
[90] *Newton's conception of space as the sensorium of God is mentioned by Addison in* The Spectator *no. 565, cited above by Wright.*
[91] *This page is an interpolation. Wright's catalogue of his manuscripts (Newcastle Public Library, Wright MSS, Vol. III) includes* ' Infinite Unity ' *under* ' Valuable Drawings in Roles '. *A description of a* ' schem ' *entitled* ' The infinite unity or sensorium of the Deity ' *(ibid.) speaks of* ' concentric parts, all rising in a destin^d ratio from an [originally: the] Ens primum ... '.

The adjacent diagram is a representation of the infinite unity, of all existing beings, or in other words the mode or image of universal creation as supposd existing in the eternal mind, not only illustrating the omnipresent state of the Deity as everywhere the same, but also shewing how the glori[ou]s presense of the Devine Nature is continually magnified with his visible creation thrô all infinity.

The temporal duration of rational beings in their several states of existence, as they pass thrô the respective regions of mortality, whither ascending towards an eternal, or descending to an enstantaneous state, evry where increase & decrease by such indifinite intervals, or fixt periods, that with regard to one another, they may all be consider'd as eternal. But nothing can be justly literally and strictly eternal but the Deity or God of Nature himself.

Thus in ye millenarie sphere making the temporal state its cube, ye duration will be 1.000.000.000 years. In that of the centuries it will [be] 1.000.000 years and in that of the decades only 1000 years. But as ye first to ye last is but as 1 to a million, the difference between a past & future state may be consider'd physically as eternal.

[3] To be eternally happy is neither absurd to expect, or unreasonable to hope for. But to be eternally happy in any one and ye same state or mode of existance is both absurd to expect, and unreasonable to hope for.

Nothing can be eternally happy in ye same state and circumstantial predicaments, but the immutable & Devine Nature itself, but beings of a finite nature may be eternally happy thrô various states & scenes of periodical enjoyments, thô it is impossible for them to be permanently so in any one, and in this will be found the true difference betwixt the Creator and his creatures, or of all visible beings, and their efficient causes: the last being subject to mortal changes, and ye other self or necessarily existing in an eternal state.

If any difficulty should arise in conceiving the indefinite denomination of creation, in its approximation to ye Punctum Stance, let it be considerd that the most enormous creation, that ye human imagination can any how form, with regard to infinity, can be no more than as a phisical point; and that all time with God is but as an eternal instant and all extent but as an infinite center.
[4[92]] I know this is very difficult to conceive, by any but profest

---

[92] *This page is probably an interpolation and the change of handwriting in mid-page suggests that its final paragraph is an afterthought.*

metaphysitians, and therefor I advise all those who are not well acquainted with such subjects to pass it over.

This diagram evedently demonstrates the sensorium of Deity not only to be an immutable unity but also infinite both with regard to magnitude and minature, and consequently its Ens Primum neither matter nor space, but somthing self existant, and as such incomprehensible to all created beings who can only reason upon the evedence of their senses.

As all these several concentric spheres of creation are distinct sensoriums of the Deity, [and] have no sensible ratio or absolute proportion to one an other, it argues very presumptivly if not positivly, that matter and space, as has been already said, can have no real existence, but as an imaginary mode in ye eternal mind; and that vertually and physically all these sensoriums of existance are infinite, i.e. ascending or infinitely increasing in magnitude, and also infinitely descending or decreasing in miniature.

[5] Hence in this most equitable light infinite mercy or justice truth &c., may be look[ed] upon as an ever flowing fountain of grace in ye ordination of temporal if not an eternal oblivion which shall render evry degree and state of temporate punishment thô far from a state of reward, yet in a due degree comparatively a state of bliss to other existences.

Justice or ye rewards of vertue, may be resolv'd into ye enlargement of ye creatures capacity, and the increase of all natural joy, not only as to ye powers of ye most grateful sensations but likewise in their eternal possession and duration.

And lastly vendiction, or the punishment of vice, not in the infliction of intolerable degrees of corporal pains but in diminution of beatitude as in grateful sensation or perhaps the increase of ungratified passions and other obnoctious circumstance of existence and situation.

[6] Thus we may not unnaturally conclude then in the great and infinite plenum of the deit[y']s refaction that the Heaven of one state or creation, may prove little more than the Hades of an other, and so on ad infinitem both ascending to infinity and descending to negation: magnitude & miniature having no proportion or distinction in ye ideas of God, God himself being only the inapproachable center & circumsphere of his own illimitable nature.

In this great chain or scale of infinite existence evry degree of happiness and misery being subordinate to ye dispensation of Providence and his eternal justice in ye awards of vertue & vice, and as subject to the infallible & consequent laws of nature, will perpetually be found most equitably subservient to ye calls of justice, mercy or vendiction, & without ye sensible agency of any corporial pains, or ye rendering of any act, event or transmutation vain, it being suppos'd to be equally within the breast & power alone of evry free agent or being whither it will voluntarily by its own vertues ascend to an angelic state, or sink by vices into a reptile or an infernal one; [7] and this perpetually by temporal changes as rising to each perfection [as is] suited to created beings, or sinking to insensibly loos all its animation thrô all infinity and all eternity, God himself as ye uncreated being only being perfect infinite & eternale.

The idea of eternal fire where ye worm dieth not and ye fire is not quenc[h]ed, is adapted only to our present sensible & mortal nature and no way suited to the wisdom of a Providence which never acts in vain as being a wast of wos[?] without benefet either to ye Divine or any other being. But if one being is promised for ye good of an other, and that the miserys of one state is destin'd to contribute to the happiness of an other, we may then venture to hope in ye language of eternal justice *all is good* and that with God *what ever is, is right*; but we must be very careful not to apply this infalible assertion to our selves, i.e. so as to vainly think that what ever Providence permits us to do is also right, (the reward being still to come).

In this we shall undoubtedly find our selves greatly mistaken, as fatally subject to infallible alternatives in ye eternal consequences of an indespensible justice.

[8] Let us suppose then an other central body with in ye orb of the Sun, and in a great degree similar in powers or operations to that without it, i.e. with a like system of planetary bodies in that internal sphere.[93] Let us also imagin if you please at ye same time an other still more vast external system of bodies as circumscribing the visible creation, the starry regions, to us being only ye internal concave of their more inormous and immeasurable Sun.

To what we know of creation indeed as reconcilable to our present set of senses & state of reasoning it may be objected that thô both these cases are equally according to nature possible & not absolutely absurd, yet are they each highly improbable since there is manifestly an almost infinite disparity betwixt ye one & ye other.

[9] To this I must beg leave to answer, that all our present conceptions are much confined & proportiond only to our

---

[93] *Cf.* W. *Whiston*, Astronomical Principles of Religion (*London, 1717*), *pp. 93-96.*

present set of senses or state of our earthly existence without ye least relation to any of the powers of a past or future state, further than to rise or excite in us suitable ideas of our dependance upon the Being of all beings in his infinite Deity or Dominion, and that in a future state all our rational faculties may be much enlarg'd, or contracted in proportion to a new state of nature, so as the infinite glory of the immutable Being may also be more or less revealed to us to answer ye awards of all degrees of vertue or of vice either in consequence of his infinite goodness or eternal justice.

[10] Here we may observe that as ye human mind in its imeterial state, has a power of creating both imaginary matter & space so as in a moment to pass from one object to another thô at an indefinite distance, so the imortal soul may also with ye like facility pass from our state of existence to an other neither of which powers at present appear to be subject to our known ideas of either mater time or space.

But God who constitutes both time & space with his infinite & eternal attributes, may also constitute these (to us at present) incomprehensible faculties.[94]

[11] The infinite and alternate mutibility of creation which constitutes all the respective changes of nature, and states of existence past, present & to come and in which all creature rewards and punishments in ye eternal despensation of justice mercy or vendiction are comprehended, may be thus illustrated and defined as

The eternal ordination of infinite mutation

Heaven	Earth	Hades
Earth	Hades	Heaven
Hades	Heaven	Earth

[12] But as the human mind is too apt to lose its self for want of more extensive powers and comprehensive faculties in the contemplation of infinite and eternal objects, it may not be amiss in some degree upon so interesting a subject to attempt as far as our finite senses & reason will allow it a more adequate idea of universal nature, and presuming as far as relates to ye present subject a definition of its mode & magnitude, we may possibly arrive at some degree of probability.

[9 cont.] Q[uod vide:] ye Chinees idea of their God & Milton's Pa[n]dem[oniu]m.

---

[94] *In the space at the foot of the page, Wright has written and then erased the word* 'Plate'.

Let us suppose, then, upon the universally admitted prinsipals of an indivisibility of matter, and of consequently an immaterial Ens Primum of all nature, that creation equally ascending and descending, magnifies & diminishes ad infinitem, nothing being mathematically to be suppos'd so small but that somthing still may be imagind less, and that nothing can be conceivd so large but that somthing [13] still may be imagind greater. And thus we shall approximate the most perfect idea of infinite magnitud & minature that can be arrived at.[95]

It must naturally & most rationally follow also from these premises that the infinite source of all wisdom can only be in possession of those perfect ideas which comprehend ye maximum & minimum of all creation, and that towards the modification of which in ye eternal mind neither solidity of matter, ye extension of space or ye duration of time may be at all necessary, the perfect unity and identity of all existences, being as ye Ens Primum of all existences in all *Punctum stans* & equally subject to an infinite Presence and an eternal now. I well know as before observ'd how difficult this is to be conceiv'd and therefore must beg leave to say further that as our ideas are no how concerned with space, substance or duration, but as by the temporal modification of our present set of senses, God's eternal ideas or maturated images of His infinite Mind, may have as little to do with ye reality of substances.

[14] The human imagination being very finite it can possibly follow the frame of nature, no farther than the powers of its conceptions will reach & allow it, our reason being tottally subject to and hence govern'd by ye sensations we in our present state are created in. And hence it is that as ye view of our present sphere of existence extends little farther into magnitude than our present telescopes will reach, or distinguish—miniture below the microscopic power of lenses—we may be justly led to allow that beings of a more divine or imaterial nature as approximating nearer to ye endowments of the immortal Mind may have sensible faculties so far superior to ours as to rise much higher in their conception of the infinite unity of all things, and also descend much lower towards the generating & prolific point of all nature. [15] From this point, an inapproachable punctum, all creation naturally proceeds with all its undulations & circulating energies as from an absolute negation of matter & consequently of motion and as occupying neither & destitute of both space and time. And thus

---

[95] *This sentence has been added.*

it phisically appears to sense & reason and next to a demonstration that universal nature agreeable to y^e *Genesis* of Moses was created out of nothing and is no other than an eternal infinite and most perfect mode of the eternal Mind.

Wh[e]ther this mode of universal creation in the perfect form of a glorious sun was ever known to ye ancients is not so easy to determine but this is certain, that somthing similar to it must have been known by ye priests of Minerva in Egypt, since as part of other misteries in sacred characters this mistic sentence was expressed:

The Sun is Born.

[16] Thus it is that by ye Devine indulgence & enlargements of our minds we may naturally arive at an almost adequate idea of the infinite unity or frame of the eternal mode, which constitutes all nature in its perfect plenum of all possibly existing beings and evry way consentanious to the acknowledg'd Devine Attributes, or infinite Wisdom and Power, or in other words the perfect image of universal nature, as perpetually existing in ye eternal Mind. That is to say, comprehending within it self thô in a manner incomprehensible to us, the creatures only of his Nature, all spirit, matter & space constituted or created out of nothing or as coeval with Him self, and as ever manifesting his Providential Presence to be evry where actually visible and self centerd in perpetual glory, always and evry where the same & immutable infinitly expanded to evry degree of magnitude and equally contracted to evry degree of miniture and in a maner as all center and all circumference and without beginning or end: the alpha of one state being ye omega of an other and one perpetual past present and to come evry where uniteing as in one eternal present moment.

[17] Thus we may rationally concive the infinite ene[r]gy of nature imbosom'd in the Deity as an image in ye human mind, vision and dream to us being only as an emanation of what in ye Divine Idea is permanent and lasting, but subject to like disolution if God or his natural laws could seece to be; for was it possible for ye Eye of Providence to close or sleep all creation would drop into its original nothingness. It[96] follows then [18] that the most rational idea we can posibly have of matter is, that it must be an eternal and infinite mode of the Divine Imagination, not existing mearly by vertue of its own innate & sollid substance only, but as ye consequent effect of our own present powers of sensation. And wh[e]ther it was created in time or co-existing always with the Deity it is all

[96] *This phrase and the page following are an interpolation.*

one to us, for we can only reason upon it according to the evedence of our present set of senses, and if our senses were eternal, mater of consequence would be so too.*

But be it what it will, as cooperating with spirit, and co-existing with space, it is manifestly subject to all the powers of mutibility and evry temporal change that Infinite Wisdom can conceive or Providence suggest as a universal good, and as such visibly constitutes the sensible and rational creation or infinite sensorium of God. [17 cont.] Thus the all infinite and universal creation may be rationally conceiv'd to be but one united emanation of ideal images flowing from ye Divine Nature, but so adapted to our senses to render them real substances and as orb circumscribing orb and center as concentering center equall inconceivable to us & inapproachable to human wisdom, both as to their infinity and definity.

The laws of nature being all ordaind or founded upon necessitous consequences it follows that they must be also subject to ye same unavoidable mutations [19] both as to time mode and place inspected of. I may have leave to say so by ye pulse of God and not only as being every where good, but as always the very best, both with regard to ye whole & in respect of the parts, and consequently immutable & coexistent with that perfection of mind which constitutes Eternal Wisdom & whose attributes can never sease to be. For if creation could be render'd better than it is, of consequence it could not be perfect. But if in consequence of Eternal Wisdom we are oblig'd to admit it all perfection, we must at the same time allow it may be also eternale.

[20] Hence with a view to the eternal welfair of the whole, there must be a sort of absolute necessity of always and evry where doing the very best and in ordaining evry where and always the best is constituted Infinite Wisdom and Goodness. Hence the laws of God and Nature must of force be immutable as perfect.

For if by a total alteration or noval disposition of ye whole frame of nature God could make his creation either better or worse, it would of course argue, either the present or ye future form an imperfection; but imperfection with regard either to his Wisdom or Power is directly repugnant to all the Divine attributes, and hence we may presume to say, as at ye end of our doxology,

> as it was in the begiñing
> is now and ever shall be
> world without end   amen.

---

* For mater perhaps in a future state, expos'd to other powers of rational conception, may possibly be destitute of evry attribut that renders it subject to our present sensation.

LETTER III

[21] *Scholia &c.*

From what has been advanc'd it may be rationally concluded that creation[97] is the eternal idea of infinite forms, universally modified by sensations, into rational and material substances. Whose indivisible unity is God, and laws of oporation Nature or Providence; and motion, generation and modification are as ye inate properties of mater. Magnitude, extension and duration are as ye same properties of space. Reason, sensation & imagination are as ye propertys of ye mind. Somthing, it is universally allow'd, must have existed from all eternity as by a fatal necessity or all things must have proceeded from nothing *which [is] a contradiction.*

In that fatal necessity is to be understood and comprehended, all ye attributes of the Deity or Divine Nature.

Therefore what ever *is* as subject to its inevitable consequences and control must infalibly be right.

It follows also as a natural consequence that this fatal necessity can not be mear matter because matter is evry where passive, but this perpetual faculty is evry where active and therefore of force we must conclude it to be a Being always & every where existing as an eternal and infinite unity co-equal in all its parts and constituting in ye whole congeries all the attributes of duration, cogitation & space.

[22] If this infinite Being or Universal Mind is self centerd, in eternal unity, all the powers of Providence & energy of Nature, flowing as from their natural minimum or Ens Primum thrô an indissoluble assembly of created or Divinely conceiv'd modes, & without any final maximum, and consequently as always & every where in perpetual essay or effect.

We have no comprehension of animal life but from our ideas of spirit, or of the existence of substance but from the evidence of our senses, but reason seems to whisper to us, that more of these sensations may be necessary to a true knowledge of God, because space, or duration, may have the same relation to ye Divine Nature as spirit has to ye human, and at the same time may be united with spirit too as we are naturally connected in our existence with external magnitude and progressive time whose essences are space & motion.

But this I say in our present state we have no reason to expect to know nor is it necessary perhaps to our salvation in a future state, God alone being the spirit of infalible trouth & unapproachable to all his cretures. [23] So that thô the nearer we arrive to infalible

---

[97] '*creation*': *the page originally began with this word.*

truth, the greater may be our title to a meritorious happiness both here and in futurity, yet unless we could unite our ideas with those of the Deity we can never presume to hope that we may be perfectly so, in an ultimate felicity, this being as inapproachably piculiar to ye eternal state of God alone and in this Divine Wisdom appears to be so just & indulgent to all its cretures, that beings in the very lowest state of ignorance by a firm and duteous adherence to ye faith or the principals of religion first imprest upon them, may be rendered as vertuously meritorious verily as the very highest, and most knowing in [c]reat[ures], the m[e]rit of faith & also of good works always sinking as a more evedent certainty arises or even probability of a reward, the moral difference lying in virtue for its own sake or a reward.

In this then consist[s] ye white unsuleable attribute of infinite mercy, as upon a vertuous supposition we may justly conclude that all human nature abandoned to their passions and in ye same disadvantageous circumstances of body and mind would from opposition, ingratitud, necessity and provocation perpetrate & suffer the same iniquity or venal offence. And this appears to have been the parient pedestale [?] of primitive Christianity.

[24 blank][25[98]] There is no possibility for the human understanding to conceive, either how the universe could be without beginning as self-existing, created out of nothing as only all ideal, or whither an eternal mode or imagery in our involuntary, or spontanious perpetuity. Yet in all these cases thô our sensations vanish in immensity & duration and our reason is quite lost, still that it was produc'd, created, or is an eternal entity of itself, is as certain and incontestable a truth as that of our own existence. And consequently thô we cannot comprehend how it is possible for our identical beings to exist in any future state, yet this is no less possible & probable than the former.

We can conceive how matter may exist thro infinite modes, & eternal changes, and if spirit was also an object of sense [26] as matter is, no doubt we should as clearly se[e] and be convinced of its immortal coexistence with it, and wh[e]ther as a constituent part of the infinite unity or an essential mode of the eternal mind, the consequences must be the same.

Let us then beleive, that in a total ignorance of this consists our greatest happiness and good, for we may well concive that if human nature were infalibly certain of all the great and awful consequences

---

[98] *A new section begins with p. 25.*

of a future state, the just and dreadful apprehension of offending the Devine Presence in evry transaction of our anxious lives, would totally deprive us of every enjoyment of our mortal beings and hence in stead of tasting, all the providential blessings destined to our natures & our present existence, render us evry moment miserable and unhappy. [27] But as the Divine Nature in all ages has reveald to the human understanding, not only evrything necessary towards the enjoyment of our present life but also what much more nearly concerns future existence, in the everlasting possession of a blest eternity. It is our duty therefore as mutual agents of his Perpetuale Providence to assist each other as much as in our power lys to improve this common happiness to evry possible degree of perfection, both as a particular & universal good.

All the evil in the world bears no more proportion to the good, than any finite dos to an infinity, or as the shade of a planet to the whole light of the system.

Besides thô an infinity can not be increasd yet any finite may be still made less and consequently thô vice may never be totally erradicated [28] yet in its consequences it may be renderd impotent & incapable of any very fatal effects.

It then naturally becomes the object of our relegion, and laws, and is the concern of all government, and as ye two first are as the sinues of the last, they aught always to be preserv'd most sacred and inviolable, and no doubt, in consiquence of an eternal rectitude, as we lessen the miserys of our fellow creatures, our own happiness will be increas'd and eternally established, according to our birthright and natural immortality.

[28′]   *Final[99] Idea*
        or Scholia ye 2^d

[29] To form some little notion of the universal law, and eternal mutability of Nature, in which mater is infinitely chang'd and modified, let us imagine Plate ye VI to represent any visible creation or present state of beings in their natural prograce from magnitude to miniture. And in this scheme ye dotted spiral line will represent the generation and process of a world from its creation or prime embrio to its final catastrophy and desolution. Here *A* is supposed ye central, *B* ye the circumorbent Sun, and *C* ye middle region or mundain space which is imagind to be divided for easy conception into seven concentric spheres.

[99] *This title is on a separate page.*

The first is the atmosphere of Heaven or of the starry firmament in which ye new world is formed, or in its first conception, as at D.

The second is ye region of material conglobation in which ye perfect sphere or globe is completed, as at E.

The third, is that of ye origin of terrafirma and of all the mineralization as at F. [30] The fourth is ye region of solar light, heat and vegitation, in which ye solar influence begins to operate with all its effects, as at G & H.

The fifth, is that of animal life and mortality in which humanity has its first existence as ye principale being, at I and K &c.

The sixth is the region of evaporation and of calcination etc. as at L.

The seventh, is ye enflam'd atmosphere of ye Sun or region of conflagration and the total solution of worlds as all consummated at M.

Thrô all those several regions a planet from its first formation revolving & round ye Sun, meeting with some resistance, thô less than any assigned quantity, falls evry revolution short of its last syzygie with ye Sun, and consequently generates by its increasing gravitation, ye spiral line of approximation to ye Sun, which we may justly call its line of temporal or distind duration or period of its existence.

[31] Thus a perpetual succession of new created worlds is or may be continually flowing from ye celestial regions to ye Sun and as finally feeding its eternal fires, and at the same time connecting all the constituent parts of creation to gether in one congeres, or infinite unity. Thus also by successive seats of new life in destind states past, present or to come and all proportioned to ye sphere of glory in which they naturally exist, all beings are inbossomed in their divine creating God.

This may be truly calld the great period of existence piculiar to evry world, and independent of each other, and in which light we may look upon ye planet Saturn in our system as not yet ariv'd at its prime perfection, the Earth to be upon its decline, and the planet Mercury to have passed its maturity so far as to be now tending to its final desolution.

This great period of nature cannot be imagind to take up less time than a million of millions of years, in which as represented in ye scheme the atmospheres of evry world are continually condensing till all their respective fluids are absorbd and finally united in ye Sun.[100]

[32] In endeavouring to account for the original of all visible beings, those arguments that are least offencive to reason and our senses aught always to be prefer'd, and no doubt are the most allowable and pleasing to ye Divine Nature. The oporations of Providence being no how mechanicall, we can have no guide to the eternall truths of God but such as are subject to ye laws of nature.

It is equall to us in our present existence wh[e]ther matter be only ideal or substantial, since ye phenomena of nature in either case & creation in general must of necessity be ye effect of infinite power & ye result of Eter[n]al Wisdom and Goodness.

If the evidence of our senses are to be taken the eternity of matter must be allow'd, but if our reason may devest itsself of such athority it may very possibly be only ideal.

In the first case it is infallibl[y] subject from ye necessity of its nature to infinite modes & changes [33] in which the Universal Mind exists by a perpetual motion and consequently [is] necessarily dependent on space.

The other as modes of ye same Universal Mind existing in ye necessity of idea only, independent of either mater or space and as being both cause & effect infinite in all his oporations, to whom no state or being is unaccountable, and illimitably happy, glorious & durable.

Hence ye first man or men may have waked into life in some most el[i]gible situation without the least knowledge of a pre-existence or how he fell asleep.

But in the first case all life must appear to vegetate as from an originating foetus of native sperm and therefore the first man & men must have been produce[d] in a different manner than by that of generating their own species.

---

[100] *A remarkable passage offering another instance of an eighteenth-century astronomer dealing in progressive change over immense periods of time. Cf. M. A. Hoskin*, William Herschel and the construction of the heavens (*London, 1963*), *pp. 68–162*; *W. Hastie*, Kant's Cosmogony, *pp. 144 seq.*

EPIGENOMA TO THE INFINITE UNITY CONCERNING THE DIVINE NATURE AND MAGNITUDE OF THE VISIBLE CREATION[101]

[1] *Of the magnitude of the visible creation*[102]

The distance of Sirius according to Hugens[103] is computed to be 27,664 radiuses of ye orbis magnus or in round numbers 276640000 diameters of ye Earth.[104] But Cassini[105] admiting the apparent diameter of the same star to 5″ and his globe equal to ye Sun, makes his distance only 3,840,000 diameters of ye Earth.

The first being equal to 2,213,120,000,000 miles
and the second ...... to 30,720,000,000.

But the largest of these Mr Bradley[106] conceivd to be much to[o] little and from many observations of ye star γ in the Dragon concluding its annual paralax to be less than 2″. This star which is of ye 3^d magnitud he afirms cannot be less than 400,000 radiuss of ye orbis magnus distant from us, or in round numbers 4,000,000,000 diameters of the Earth & equal to 32,000,000,000,000 miles. This Whiston[107] from ye observation of Mr Framsted [sic.] & Hook who determined the parallax of the Pole Star, ye 1st: 45″, ye 2^d 50″ concludes its distance at a mean of 47″ to be no more than 9000 radiuss of the orbis magnus or in round numbers 90,000,000 diameters of ye Earth & equal to 720,000,000,000 miles. [2] It is evedent from the forgoing numbers, that if the solar system was twice as large as it is, there is space enough according to Cassini, for 10 such systems as ours[108] and if we double

---

[101] *Once again we have a patchwork, though this time the pieces are difficult to separate. The original title page probably read* 'Epigenoma concerning the Divine Nature', *and after inserting the additional material Wright has forgotten to supply the original manuscript, which now begins on p. 9, with a separate subtitle.*

[102] *The original brief title is illegible.*

[103] C. Huygens, Cosmotheoros (*Hagae-Comitum, 1698*), pp. 136–7.

[104] *The diameter of the Earth was known to be a little under 8000 miles* (cf. Clavis Cœlestis, *p. 5*), *so Wright is taking the radius of the orbis magnus as about eighty million miles.*

[105] *Jacques Cassini,* 'De la grandeur des étoiles fixes et de leur distance à la Terre', Histoire de l'Académie royale des sciences, année MDCCXVII, avec les mémoires (*Paris, 1741*), Memoires *pp. 256–68, p. 259; cf. E. Halley,* 'Some remarks on a late essay of Mr. Cassini', Philosophical Transactions *31 (1720–21), pp. 1–4.*

[106] *James Bradley,* 'An account of a new discovered motion of the fixed stars', Philosophical Transactions *35 (1727–28), pp. 637–61.*

[107] Cf. W. Whiston, Praelectiones Astronomicae (*Cambridge, 1707*), IV.

[108] *Wright has jotted some of the calculations in the margins. In* Clavis Cœlestis, *p. 4, he gives the mean distance of Saturn from the Sun to be 777 million miles, so that the diameter of a planetary system twice as large as ours (ignoring comets (cf. p. [5] below) and possible trans-Saturnian planets) would be over 3,000,000,000 miles.*

the radius of such a system there will be still space enough left for ye orbits of 5 of them betwixt Sirius & ye Sun and room for 100 such systems of stars in the whole plain and to 1000 in the whole spherical aria: this number yet is much increased by Hugens; in the first instance to 712, in the 2^d to 506,944 and in the last to 360,944,128. Whiston^s computation fals much short of these, as in first instance it is only 232, in the 2^d 51,824 and in the last 12,423,168. But in some of the stars[109] particularly γ Draconis, these vast numbers are astonishingly increased *vide Bradley* for in ye 1st instance it is 10,210, in ye 2^d: 104,244,100 and in the third will be no less than* 1,064,332,261,000. But this is in ye whole sphere of Heaven & supposing the solar stars had no motion.[110] [3] Of all this increadable number w[e] have not yet observed the correct places of more than 3000,[111] but besides the above which are all supposd to be either fix'd or erratic in ye mundain space & without which the immeasurable area of the visible creation, would be little less than a vacuum & all un occupied, there are infinite others fix't in ye solid firmament of heaven which have in every respect a like appearance, but are in reality no other than vast ignovomous fountains of etherial fire, or inflamable matters, ordaind by the Infinite Wisdom of God, no doubt, to feed the numerous Sun or solar bodies with perpetual fuel.[112] But still in some measure to lessen the prodigious number of intermediate bodies, we may well imagine that many of the planetary systems are remov'd a very considerable distance from the combus[t]ible firmament of Heaven in which there appears to be a perpetual war of elements no less than may be justly imagind in the great limb[us] of Nature [4] from whence the Divine Energy produces the creation. We may also as rationally conclude that there is an awful distance from the inner orbits of the visible creation to ye centeral

---

* In all the estimation of ye magnitude or distance of ye stars, no astronomer has ever yet presum'd to aim at precision, so that we must content ourselves with such conjectors as will best satisfie our reason.

---

[109] *Wright is prepared to believe the results of both Bradley and Hooke, although Bradley's work was a devastating rescrutiny of Hooke's claim to have measured the parallax of Gamma Draconis.*

[110] *Bradley explains that proper motions of the Sun and/or of the observed stars will affect the measurements of annual parallax, 'An apparent motion observed in some of the fixed stars', Philosophical Transactions, 45 (1748), pp. 1–43, espec. p. 40.*

[111] *In John Flamsteed,* Historia Cœlestis Britannica *(3 vols., London, 1725).*

[112] *Cf. Letter I, p. 34.*

globe, or more excellent body from whence the whole is govern'd, but of this great central sphere or ratiorum, I have spoke in other places,[113] so at present I shall conclude this most interesting & sublime subject, with my ardent wishes, that it may wake some more able genius to consider it much further & I hope to ye approbation of the more learned.

Plate the VII represents an idea of a creation of this construction, in which the central body is of a terrestrial nature and not a luminous as our Sun but with a system of suns instead of dark bodies moving round it & to be further considered by astronomers.[114]

[5] I know that it may be objected and not without some reason, that the solar system is here conjectur'd to be much to[o] small, especially as the comet of 1680 in the Newtonian astronomy is carried 14 times farther than the orbit of Saturn from the Sun.[115]

But here I am constrain'd to observe that notwithstanding the allow'd abilities of that great philosopher the present theory of the comets may possibly be very fallacious for I can't help thinking, as I have indeavourd to prove, in Plate III,[116] that the comets proximity to ye Sun and its vast aphelion distance are comparatively too preposterous to be easily admitted for in all our conclusions, we aught to be very cautious not too obviously to offend our reason or insult the human understanding.* [6] We may then most rationally conclude that all new or extinct stars, as also most of the variable and cloudy stars, are all of an ignivomous nature and most probably in the solled frame or firmament of Heaven, consequently no other than celestial vulcanos, bursting from or dying a way in their various *craters*. But such stars as gradually increase, or diminish their brightness or luster, may like our Sun become more and more obscure from their own fuliginous eruptions & maculae, or other wise, grow more and more brilliant from new combustible

---

* The regions in which comets have yet only appeard are that of Bootes, Orion, Leo, Lyra, Syrius and Aquilla and Regulus.

---

[113] *In Letter IX of the* Original Theory. *Wright is now supposing that our Sun, with certain stars, is in orbit about the true centre of the system, or else fixed at some distance from the centre, a conception closer to his 1734 and 1750 positions than the rest of the present work.*
[114] *Very probably a later interpolation.*
[115] *Cf. W.* Whiston, Astronomical Principles of Religion (*London, 1717*), p. 25.
[116] *In Plate III: an interpolation.*

explosions.[117] These we may conceive to have like our Sun, planetary systems around them. Of these as the brightest & nearest to our system we may recon

   Capella
   Lyra
   Aldebaran
   Shoulder of Orion ..... 1° more South
   Lyons Heart
   Spice Virgenes
   Sirius ..... 15' more North
   Arcturus ..... 7° Latitud.*

Some of which have been observed[118] to have local motion, as p[e]r Morgan.[119] [7] As to what Mr Bradley says[120] upon the subject, if it tends to prove any thing it is that the stars are not exactly seen in their true spherical places; but from the perfect parallism of the tube thrô which his particle of light is imagind to pass, in its progressive motion to ye Earth, it is very evedent, that all the stars must still appear in ye same intangible [?] points, since every particle of light, by which they are visible, & successively ariving at ye eye, never change their direction. Besides, if ye aberration of ye stars were really occasiond by ye progressive motion of light, no very satisfactory reason can be given, why one star should not be more effected than others by it in ye same line of direction as [it] in consequence of their very different and various distances. Besides as** Mr Bradley is willing him self to believe, that ye radius of ye orbis magnus to ye least distance of ye nearest star is but as 1 to 400,000; in this case it may be considerd only as a phisical point,

 * or nearly 1° in 1000 years.
 ** See Long's *Astronomy* p. 302.

---

  [117] *Newton supposed stars 'may be recruited by comets that fall upon them; and from this fresh supply of new fuel these old stars, acquiring new splendor, may pass for new stars', while variable stars may by their rotation present to us alternately their light and dark sides. Cf.* Principia, *Motte-Cajori translation, pp. 541–2.*
  [118] *E. Halley, 'Considerations on the change of the latitudes of some of the principal fixt stars',* Philosophical Transactions *30 (1717–19), pp. 736–8. The angles Wright has jotted down probably refer to these changes.*
  [119] *Not identified. George Cadogan Morgan (1754–98) was a scientific writer of repute but this section of Wright's manuscript is probably too early for him to be the 'Morgan' cited.*
  [120] *J. Bradley, 'An account of a new discovered motion of the fixed stars'. Wright confesses his inability to understand Bradley's conclusions; in particular, he does not appreciate that what matters is the speed of the light reaching the observer, not the distance it has travelled. But since Wright is depending upon Long's laborious treatment, there is some excuse for him.*

to which in an other place he affirms the stars can have no aberration. But gentlemen of more sagasity & greater penetration [?] may perhaps be able to conceive what the writer cannot possibly comprehend.

[8[121]] In this hypotheses, the Via Lactea is lookd upon as no other than a vast chain of burning mountains forming a flood of fire surrounding the whole starry regions, and no how different from other luminous spaces, but in ye number of stars that compose them, or where there are none, in the vast floods of celestial lava that form it. In many of which for a certain time, some brighter parts may appear, such as ye maculae[122] or fiacula[123] in the Sun. And such have also appeared in the floods of lava issuing from vulcanos upon Earth. See Lord *Winchelseas* account of *Etna* when ambasada at y^e *port*.[124]

The idea of a sollid firmament is no new hypotheses, since it was first, & always concivd to be so by ye ancients, especially by Pythagoras, Ptolemy &c.

[9]   [*Of the Divine Nature*]

That no finite nature can ever comprehend the attributes of an infinite one is evedent to a demonstration from what follows.

All our senses and consequently our reason, are in our present natures limited, and as confin'd to the present state of our human existence, and also proba[b]ly created, for no other use than the proper business and sole end of our mortal life, and as the sense of sight & also that of hearing being both finite, the imagination which always forms it self upon ye objects of sensation only, must also be finite. Hence it is no wonder that we cant possibly conceive all the attributes of the Divine Nature in their full extent, especially since as being all of an infinite nature, there can be no end to our imaginations in the search for them. For as in sight there is still something to be conceiv'd beyond all visible objects which we can not possibly perceive, as also in sounds a distant harmony, still more remote & what we can't possibly hear, so there must like wise be and as rationally to be concluded in every state of existance [10] somthing remaining to be still known, and consequently in ye all excellent nature of the Deaty, much to be concived which we can not possibly

---

[121] *The material on this page, which is the verso of p. 7, may be a later addition.*
[122] '*Maculae*': *sunspots (Latin: ' spots ')*.
[123] '*fiacula*' [recte '*faculae*']: *bright areas on the surface of the Sun (Latin: from ' fax ', ' torch ')*.
[124] *Earle of Winchilsea, A true and exact relation of the late prodigious earthquake and eruption of Mount Ætna (London, 1669), 38 pp., 1 plate.*

expect to comprehend. And therefore whither God is the all governing Spirit or of a Divine material Nature, 'tis of equal consequence to us, since as infinite He can never be represented, as an object of sense and consequently can never be comprehended by any finite or created being.

But thô a perfect knowledge of the Deaty in our present state, and probably in no other, may not absolutely be requir'd, yet a sufficient knowledge or consciousness, of this great Author of our being may be indispensabl[y] necessary, so far as to establish our faith in his Providence, and a future existence. The consequence of which, as resulting from our ideas of his eternal justice, and goodness, must as infalibly be ye most equitable rewards, and punishments in a future state.

[11] We perceive space, we concive time, and we are conscious of spirit, but all these have only relation to and constitute the immaterial attributes of the Deaty, to us they are objects only of the mind or understanding.

But we have a sensible demonstration of mater as the object of our reason, in the mode of our existance with all the other sensations of life.

Hence it is that by the first of those ideas we can only arrive at such notion or conception of a god, as ye eternal Creator of all things and by the latter that we can possibly attain to any adequate knowledge of our selves, either as his creatures, & created by him, or as proceeding from him, and hence of course our immortality.

The result of what has been said will naturally lead us to prove that mater is only ye manifestation of a god, without which we could have no conception of a deity. But that in reality as a solid substance [it] has either no existence but in sensation, or must be itself eternal. Hence to us it is to all intents & purposes, a real corporial substance, but to ye Divine Nature whose existence is immaterial and neither subject to or ye object of sensation to any of its creatures, it may be only [12] as an image, and such as are involuntarily existed in our dreams, but in the Devine Nature the act or fiat of his will producing creation. Thus with relation to spirit, we may infer that there is neither time nor space, but with regard to mater as the object of sense, there must be boath.

That space is God, or was looked upon by the primative Christians as the father of all beings in his infinite censorium, we have from these lines in St Paul:

In Him we live & move and have our being.[125]

---

[125] *Acts 17 : 28.*

That duration or time was also imagined to be God appears from his being called the *ancient of days*.   Daniel 7 : 9

That spirit has been likewise judged to be God we may learn from y^e *Genesis* of Moses:

    As the Spirit of God moved upon the face of the waters.[126]

Again:

    He breathed into his nostrils the breath of life.[127]

But mater is no where said to be God, yet matter is manifestly ye sensorium of spirit as space is visibly ye sensorium of matter & time most evedently is ye eternal sensorium of all duration as eternity is the sensorium of time.

[13] As the same common Earth produces all the great variety of vegitation, and vegitation producing in effect all the parts and variegation of animal existence it seems sufficiently to evince that all the primitive particles of matter are homogenius and as ye same. But if the origin of all beings of ye same species being only ye same it will follow that every particle of ye same creature in its ens primum will produce ye same or analogus beings, and consequently also an infinity of them. This would intimate, or lead the ancients to concive, that God & Nature were ye same.

But matter, at least created matter, is not God, because matter is subject to motion, and circulation, but God is every where without motion, and knows everything without reflection or thought, not as all human minds from reason & contemplation, which is progressive from one object or one sensation to an other, but as being in all things, immutable and as every where consciousness its self.* [14] The great *quere* and in which we are much concernd is whither human nature is capable of reviving again in this world or only subject to a resurrection upon ye final desolution of ye present frame of nature. But this is a peace of knowledge in our present state we can never expect to arive at but by revelation, and as the latter seems to be most rational[ly] adopted in our present religious faith. Yet when we reflect, that to all insensible beings, an eternity is but as a moment, we nead not be very anxious about the real trouth.

[15] The ancient Persians conceiv'd that the principals of good and evil were like light and shade upon Earth, nearly equal, and in consequence of that belief worshipped both a Good and Eval

---

    * Hence St Paul has: In Him we live and move and have our being.

---

[126] *Genesis 1 : 2.*
[127] *Genesis 2 : 7.*

Genius, which was their Arimaos and Arimandes. But in this they were very erronious, for in truth good and evil bear no more proportion to one an other in this world, than as the pangs of death to ye joys of life. Some few penal circumstances & situation no doubt their are, but in general, amongst the indefinite numbers of human nature, nothing but absolute id[l]eness is subject to misery, the Devine Nature its self being always active; so that in ye white hopes of eternal truth & justice, our state of probation is very bearable.

Now rewards and happiness we can well conceive: but what is punishment?

Infinite power can do nothing in vain, therefore it is very natural to conclude [16] that the punishment of one creature is ordaind for the rewards or benefit of another and hence if instead of eternal flames, where we are told *the worm dyeth not and the fire is not quenched*, we substitute a state of inferior natures, as the awards of vice, we shall then see clearly that all superior states and natures must be ye infalible consequence of & the rewards of virtue—constituting the Hierarchy of Heaven but also that of Hades. But of this more in its proper place.

[16'] Timotheus wrote that Orpheus, who manefestly drew his doctrine from the Egyptians, [taught]:

That all things were made or produc'd by one God-head, or Divine Nature, of three names; and that this God or his infinite unity was all things, or Nature, in universal & eternal being.

And that these three incomprehensible beings or names had been further explained to be* Counsel, Light, and the Giver of Life; and as nothing can act where it is not, or in an inconscious vacuum, these three most significant terms, perhaps may be better understood by others nearly sinonimus, i.e. by the words spirit, matter, and space, in which sense tho all ye 3 are distinct existences & from all eternity, yet in their union or infinite connection, they constitute but one sensorium or eternal state of the Divine Nature. Upon this probably St Athenasus, who was an Egyptian & no doubt w[e]ll skild in his native doctrines, form'd his Chris[tia]n creed.

[17] Thus a metaphysician might justly form a paraphrase upon St Athanasius' Creed.** The most rational faith is this, that we believe one Divine Nature in a trinity of attributes, and the trinity

---

* To the *Aether* and *Chaos* of Orpheus, according to Syriacus, Simplicius adds *Time*.
[16 cont.] ** St Athenasius was not only him self an Egyptian, but also may be suppos'd well acquainted with all their secret learning & religious tenets.

of attributes in a unity of the God Head, neither confounding their attributes or dividing their co-existence. For there is one idea of existence of time, an other of spirit, and an other of space, but the attributes of space, spirit, & time are all equal and co eternal. Such as space is, such is spirit, and such is time, all uncreated, i.e. space uncreated, spirit uncreated and time uncreated; all likewise incomprehensible, i.e. space incomprehensible, spirit incomprehensible, and time incomprehensible. Space is also eternal, and time or duration eternal.

But yet there are not three eternal God Heads but one eternal Divine Nature, as also there are not three incomprehensible God Heads, nor three uncreated Divine Natures. But one uncreated Divine Nature & one incomprehensible God Head. [18] So space is all, or infinitely mighty, spirit almighty, and time almighty, but yet there are not three independent almighties, but one co-existent Almighty. Hence spirit is God, space is God and duration is God, but yet there are not three Gods, but one eternal union or God Head: for no one can exist without the others. All these are made of none, neither created not begotten, and as their is but one space, one spirit and one duration, none is afore or after others, none is greater or less than another. But the whole three beings, existences or attributes, are infinitely co-equal, and co-eternal and constitute one infinite unity & eternal Godhead, and in ye trinity of that unity, is only to be comprehended all existence & every created being. Thus and if matter is substituted in ye place of Athanasus['s] son it will follow that matter is of space alone neither made nor created but begotten.

[19] Death or mortality is evedently spirit deprived of all sensation, after which all ye members of the body are held together only by a natural cohesion, gravity or attractions of ye constituent parts. But it does not follow that when they are totally seperated and dispersed, that they are absolutely non entities and as destitute of an ens primum incapable of being restord to life again. For thô they appear to be no longer active, yet may they be still the passive agents of infinite power, and always ready for a renovation when time and Providence shall call them forth.

Hence we may be lead to beleive that every particle of matter may be endow'd with spirit, thô totally, for want of ye organs of sensation, in[s]ensible of any situation it may fall into as to pain or pleasure, as being lock'd up in apathy or oblivion. Yet as its

[18 cont.] The Sun the visible throne of ye Divine Nature see Daniel 7: 9 and 10.

ens primum or identical existence is not destroyed or taken away [20] when ever it falls, or arives into a state [of] regeneration or of homogenious nutriment as in ye simini animaly[128] or of its own homeomery, it may as of an eternal necessity be restor'd to life.

Thus the primogenial attoms of one body may possibly become parts of any other, without being ye least effected by them, but when they become part of their own original natures again in the unity of a foetus, they again flow into life and become sensible both of duration & space.

Thus if the homeomery of all the primordial particles of life as unisons in music, produce the same being or species of animals, and also all the variety of annimal beings, from the different organs of sensation, as the effect of circumstance or situation, these must all be directed by a Providence or we shut out wisdom & power, which is most manefest from the beautiful productions of creation in general.

[21] But then from the same evedence, it necessari[l]y follows that he is all powerful, all knowing, and all wise, and of this the beauty, order, and disposition of ye creation, with ye infinite variety and temporal regularity of its parts is an absolute demonstration; and it is equal to us in the final consequence wh[e]ther the trinity or these tremendous powers is comprehended in ye unity of a Divine Nature, or the unity of the Devine Nature is contained in ye trinity of these self existing attributes.

It is also much the same to us wh[e]ther mater is self existent, as an eternal necessary being, or only a consequent and imaginary mode of ye uncreated & eternal Mind. For both in its active and its passive state it manefestly has all ye excellences and attributes of ye Deaty or Devine Nature it self. Vid: It has space in its extension & magnitude, [22] it has time or duration in it[s] modes of motion and it has spirit in its inspirations and sensations or conciousness, or other wise in its attractions cohesions & mutual gravity. So there appears to be nothing lost in supposing it may be part of ye essense of Deaty.

But then God or the Devine Nature is every where existing and ye same in equall dignity & state thrô all creation. But mater is not so, mater visibly exists in infinite variety of grad[at]ions, and consequently thô equally perfect, not equally excellent, & hence it

---

[128] *Probably Wright intends ' semina animalia '.*

cannot be God, as the will of anything is not the thing, nor ye active or effective power of any thing the cause.

[23] But again, if matter has all the excellences of the Divine Nature in it, as subject to neither real corruption or anihilation and as reducible from any state by regeneration to a state of greater perfection, it may also have the powers of supream excellence in it also and consequently every possible degree of perfection in its progression to an infinite and eternal state of glory.

Where then is the difference to any mortal being wh[e]ther it was created from nothing, or as an emanation from ye infinite essence of God, since ye same power that produced it still evidently exists, and can again restore it to another life.

Those who would make God a self existent, necessary and involuntary being seem to forget, or omit, his infinite wisdom & apparent Providence, as also his infinite power and goodness. Give him but these vertues also, as absolute necessities at ye same time, and let us call him what we please. [24] He is still the same incomprehensible God in all sensoriums, i.e. wh[e]ther as existing from an absolute necessity, in spirit space & time, or in eternal matter, and thô the infalible trouth of all this can never be reasonabl[y] expected, yet it is enough in our present state to establish in us a principal of vertue, justice & goodness & as ye mutual agents of his Providence to one an other which is ye sole object and great end of all religion.

[25] In the eight[h] vol^m of the *Spectators*, no. 600, the very philosophical writer[129] is inclin'd to believe that every desire of ye happy will, in a future state, be follow'd by immediate fruition, that is, " If we wish to be in groves or bowers amongst running streams or falls of water, we shall immediate find our selfs in the midst of such a scene, as we desire; if we would be entertaind with musick and the melody of sounds, the consort will rise upon our wish and the whole region about us be fill'd with harmony, &c." This even is not more miraculus than any of ye visible productions of nature. But this will by no means agree with our ideas as created beings; for by the same supposition a finite being may most absurdly wish to be an infinite one which can only suit with the Devine Nature. But this notion of unlimited happiness suits very well with ye attributes of infinite power, and we may well imagin that all such wills of the Devine Nature and every creation its self, may perpetually exist in it, and as regular & periodical, and hence may arise

---

[129] *Addison. No. 600 is among issues of* The Spectator *noted on the opening page of Wright's Preface above.*

in a providential maner all the infinite changes of nature as summer winter, day night, activity & rest, in an eternal state of circulation. [26] Thus if the Devine Nature can be concived to will its own instantanious happiness, and to every unlimite[d] degree of perfect satisfaction immediate fruition must also be ye consequence, since to wish & will with infinite power are sinonimus terms. Consequently this incomprehensible Being must ever be in the fullest possession of every rational or sensible joy, and as this great Creator is both ye author of sensation and reason, he may by his infalible & divine justice admit all worthy beings to any finite degree of this eternal fillicity; but the infinite capacity, like infallible truth, can only be centerd in him self.

[27] If the joys of heaven are perfect and eternal, in heaven there can be no existance of that most excellent of all vertues, hope; for the same reason, if the infinity of the Devine Nature were its self, and not every where in its creatures, it must of necessity be totally devested of this great soul enriching principal and consequently more subject to misery than any of its dependant beings, which is in ye last degree absurd.

There fore when we attempt to form any right notion of the devine powers, we must not only imagin the creation immutible, but also infinitely extended, and in which all the attributes of the Deity are perpetually existed, both in expantion & contraction and from perfection to perfection, both in its magnitude and miniture, to whose final consummation it can never reach, being as every where ending and every where beginning, or as a ray of light in perpetual progression without any material obstruction, can never be stopd. But as a person born blind can have no idea of light or colours, so for want of some unknown faculty in our imaginations, we are render'd in our present state of existence totally incapable of forming any just or adequate idea of a Nature who is only comprehended in infinity of which we have no other notion than it is a Being upon which all other beings depend. *

[28] But if one of the principal attributes of the Devine Nature in consequence of his eternal & infinite existence be activity, rest as equally must be an annihilation to it; therefore mut[a]bility in its internal state or economy must be as necessary to its Providence as any other of his eternal attributes, for as human nature finds its self relieved by various scenes and changes of amusement, the Infinite

[27 cont.] Q: N° 600 S[pectator].

Nature may also require infinite modes of like mutation to supp[l]y all its infinite faculties with subject matter for his almighty power to work upon.

[29] *Corollary*

If spirit can exist without matter, spirit may also exist without space; but Spirit cant exist without idea, consequently spirit, idea & space naturally constitutes a general or œcuminical unity, and if for ye generative term *idea* we substitute its genus *matter*, in matter space & spirit, all creation will be comprehended, as an infinite & eternal unity and mode of ye universal mind.

Again, if matter was generated or created as a solid and real substance, there must also have been a time & space without creation, and hence all nature will be reduced to its original spirit or idea; so that what ever is ye case, ye infinite and eternal consequences are ye same.

Hence also it evedantly follows that God is not ye world or universal creation, any more than as idea can not be spirit or its own existing principle. Spirit in human natures we must allow is some times without ideas, and I know no reason why there may not come a time when all creation may for a season also seace and Providence may rest from all its works as in a state of apathy or as in an indefinite Saboth. But still [35^{130}] it will be the eternal Providence with all its energy & faculties, concentering in its self. For thô universal nature is eternal & can never be distroy'd, its perpetual laws or necessary consequences may be at length a little relaxed as all the circulating season[s] upon Earth are in vigorated and recruted by a distant winter and even ye spirit & human nature its self is reviv'd by rest & sleep: besides, it would be assuming to[o] much as creatures to suppose any state to be identically eternal but that of God or the Creator him self.

To object to any part of this will be in some measure to deney both the progression of time, and motion; for we are taught to believe that in ye beginning God created the heavens & ye Earth &c., and also rested on the seventh day. But if we allow of this universal Saboth, and an infinite vecessitud of life & eternal seasons, God will only have ye same relation to universal nature as the soul of man to his own corporial frame, and in this most significantly man may justly be said to be created in Gods own image, [36] and the œcumenical creation no other than an infinite and eternal mode

---

[130] *There are no pages numbered 30–34; p. 35 is the verso of p. 29.*

of his universal mind.   I[131] shall conclude with a sentence out of a
very eminent poet & philosopher, put into ye mouth of one of y^e
most illustrious Romans:

If there is a power above us, and that there is all Nature has
allow'd thrô all her works, he must delight in vertue,
and that which he delights in must be happy.

---

[131] *This sentence originally read:* '*Hear let us hold.*'

INDEX TO THE PLATES

Plates.

Letter ye I.

I   An Illustration of Natural Phenomena demonstrating the principals of the hypothesis.

Page 22 [19, 30]

II The New Laws of Comets & New Stars Explaind.

[Pp. 21–23, 30]

III Sr Isaac Newtons Theory of Comets Refuted.

Page 31 [29, 31′–31″; Epigenoma p. 5]

Letter II.

IIII The Visible Creation Illustrated, in which *A* represents ye supposition of a superior sun with out ye starry firmament & *B* the conjecture of a new super celestial earth.

Page 3^d [10; Letter I p. 33]

Letter III.

V   An Idea of the Infinite Unity or the existing mode of all created beings.

Page 2^d [–4; Letter II p. 11]

VI An Illustration of ye Creation, Generation & Desolution of all Worlds.

Page 29 [30]

Epigenoma.

VII An Idea of a Celestial Earth in the center of a sun, or in an orb of fire with solar planets round it, as in Plate IIII at *A*. The Idea of a Super Celestial Creation in which all the visible Universe constitutes but one of its planetary bodies as supposed at *B* in Plate IV.

[P. 4]

INDEX OF NAMES

Adam, 24.
Addison, Joseph, 19, 48, 61, 85.
Anaxagoras, 23.
Anaximenes, 23.
Apion, 59.
Aratus, 22, 23.
Ath[a]nasius, St., 82, 83.

Baker, Henry, 24.
Bevis, John, 33.
Bode, Johann Elert, 43.
Bradley, James, 33, 56, 57, 75, 76, 78.
Burnet, Thomas, 27.

Caesar, Germanicus, 23.
Cassini, Jacques, 75.
Cassini, Jean Dominique, 43.
Cicero, 23.
Clairaut, Alexis Claude, 26.

Daniel, 81, 83.
Day, Mark, 33.
Desidets, 58.
Dingle, Herbert, 7.
Diogenes, 23
Docetists, 58.
Doyle, A. I., 15.
Dryden, John, 26.

Einstein, Albert, 50.
Emped[o]cles, 23.

Flamsteed, John, 57, 75, 76.
Fludd, Robert, 51.

Gorgons, 21.
Graiae [Graeae], 21.
Gregory, David, 56.
Grosser, Morton, 43.
Gushee, Vera, 9.

Halley, Edmond, 10, 13, 26, 27, 31, 34, 78.
Hamilton, William, 32.
Hastie, William, 7, 9, 45, 73.
Heraclitus, 58.
Herschel, William, 43, 45, 73.
Hipparchus, 34.
Hooke, Robert, 57, 75, 76.
Hornsby, Thomas, 56.
Hughes, Edward, 7.
Huygens, Christiaan, 34, 75.

Isaiah, 47.

John, St., 53.
Josephus, Flavius, 52, 53, 59.
Jouties, 58.

Kant, Immanuel, 9, 45, 73.
Kepler, Johann, 35, 40, 42, 43.
Knight, D. M., 8.

Lambert, Johann Heinrich, 7.
Lavros, Marquis of, 32.
Locke, John, 20.
Long, Roger, 34, 35, 43, 78.
Lovejoy, Arthur O., 20.
Lycurgus, 52.

Manilius, Marcus, 52.
Martin, G. R., 7.
Maupertuis, Pierre Louis Moreau de, 9.
Michell, John, 33.
Milton, John, 46, 52, 53, 54, 65.
Morgan, 78.
Morgan, George Cadogan, 78.
Molyneux, Samuel, 57.
Moses, 81.
Munckley, Nicholas, 33.

Needham, Joseph, 24.
Newton, Isaac, 19, 26, 27, 29, 30, 31, 33, 37, 39, 41, 53, 55, 56, 61, 77, 78, 89.

INDEX OF NAMES

Olbers, Wilhelm, 43.
Orpheus, 82.

Pagel, Walter, 51.
Paneth, Eva, 7.
Paneth, F. A., 7.
Paul, St., 23, 47, 52, 53, 80, 81.
Pericles, 23.
Perseus, 21.
Piazzi, Giuseppe, 43.
Plato, 52, 59.
Plutarch, 23.
Prometheus, 21.
Ptolemy, 34, 79.
Pygmalion, 21.
Pythagoras, 58, 79.

Rafinesque, C. S., 7.

Sagasities, 21.
Senex, John, 39.
Sherburne, Edward, 52.
Short, James, 33.
Simplicius, 82.
Solon, 52.
Suffis, 58.
Swinden, Tobias, 45.
Syriacus, 82.

Timotheus, 82.
Titius [Tietz], Johann Daniel, 42.

Virgil, 58.
Vulcan (myth.), 21.

Wall, William, 45.
Whiston, William, 10, 12, 27, 28, 29, 32, 39, 44, 45, 51, 53, 57, 64, 75, 76, 77.
Winchilsea, Earl of, 79.
Woolf, Harry, 56.

Zeno, 58.

B
2